Sculpting Concrete Garden Ornaments

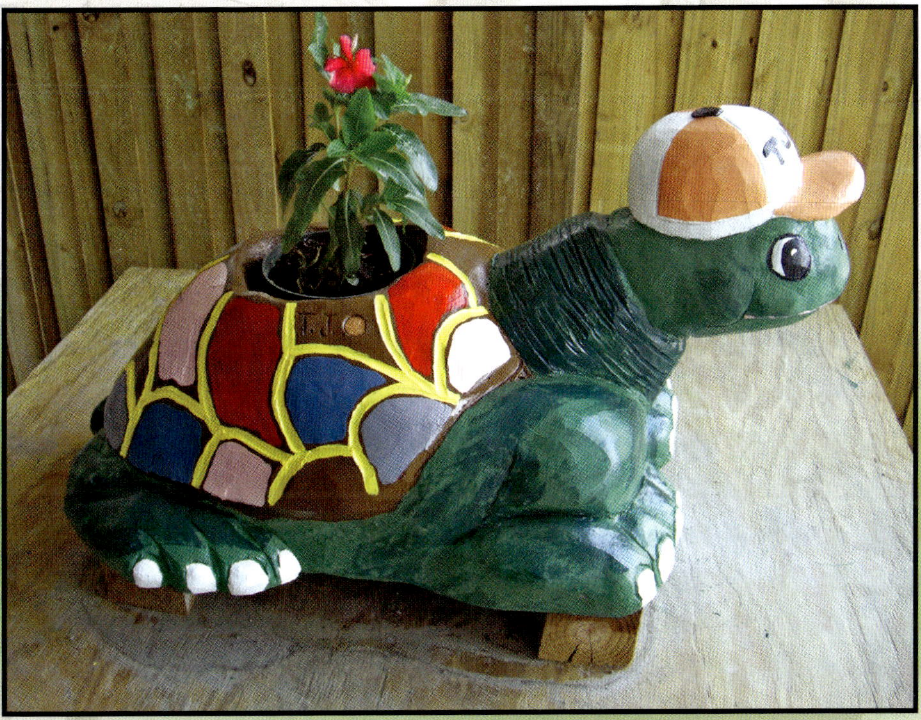

T.J. Neil

Schiffer Publishing Ltd

4880 Lower Valley Road, Atglen, Pennsylvania 19310

Acknowledgments

The following people helped make this book possible. I thank you all!
My wife and girlfriend of forty-six years, Jean T. Neil, for all the typing, work and (patience with me);
My son, T.J. Neil, Jr. Thanks for the great photography work and artistic input;
My daughter, Tina L. Hayes, for her massive input.

Contents

Other Schiffer Books by T.J. Neil
Making Concrete Sculpture.

Other Schiffer Books on Related Subjects
Arts & Crafts Era Concrete Projects. Pedro J. Lemos & Reta A. Lemos.
Backyards and Boulevards. David R. Smith & Interlocking Concrete Pavement Institute.
Creative Patios. Tina Skinner.
Decorating with Concrete Outdoors: Driveways, Paths & Patios, Pool Decks, & More. Tina Skinner & ConcreteNetwork.com
Designs for Garden Paths. Heidi Howcroft.
Hardscaping with Decorative Concrete. Tina Skinner.
Patios, Driveways, and Plazas: The Pattern Language of Concrete Pavers. David R. Smith & Interlocking Concrete Pavement Institute.
Paver Projects: Designs for Amazing Outdoor Environments. Melissa Cardona & PaverModule.

Library of Congress Control Number: 2007942167

Designed by Mark David Bowyer
Type set in MarkerFeltWide / Humanist 521 BT

ISBN: 978-0-7643-2875-6
Printed in China

Schiffer Books are available at special discounts for bulk purchases for sales promotions or premiums. Special editions, including personalized covers, corporate imprints, and excerpts can be created in large quantities for special needs. For more information contact the publisher:

Published by Schiffer Publishing Ltd.
4880 Lower Valley Road
Atglen, PA 19310
Phone: (610) 593-1777; Fax: (610) 593-2002
E-mail: Info@schifferbooks.com

For the largest selection of fine reference books on this and related subjects, please visit our web site at
www.schifferbooks.com
We are always looking for people to write books on new and related subjects.
If you have an idea for a book please contact us at the above address.

This book may be purchased from the publisher.
Include $3.95 for shipping.
Please try your bookstore first.
You may write for a free catalog.

In Europe, Schiffer books are distributed by
Bushwood Books
6 Marksbury Ave.
Kew Gardens
Surrey TW9 4JF England
Phone: 44 (0) 20 8392-8585; Fax: 44 (0) 20 8392-9876
E-mail: info@bushwoodbooks.co.uk
Website: www.bushwoodbooks.co.uk
Free postage in the U.K., Europe; air mail at cost.

Introduction

For these four small sculptures I have drawn up rough sketches as guides. They may be seen in the photographs. From these you will be able to create original works of art. However, I think it would be better for you to try creating the first sculpture using a small model (statue) of some kind you may have around your house for reference. I believe a three-dimensional statue will help you a great deal; however, always remember that it is not a big deal if you make a mistake. It does not matter much if, lets say, one foot is a little bigger than the other. The whole object should be to have fun with your project. Once you have mastered the techniques described here, you will soon create a sculpture you will be proud to sign with your initials. Maybe you will also pick up on my tradition of adding a penny to the concrete for luck.

I will be using the term "**mud**" in place of the word "concrete" throughout the book because I'm old school! If you mix Portland cement, sand, and small stone together, that's what makes concrete. In this case, we are replacing stone with galvanized lathe wire for strength. These four sculptures do not require any rebar; they will be quite strong with just lathe wire. With small statues, you don't have to order materials by the ton. Go to your local building supply and get ready mix in bags. When you buy lathe wire, it comes in bundles of 10 sheets, 27" wide and 8' long. Most building supply stores will allow you to buy one sheet at a time. The lathe you want for these sculptures should be 2.5 gauge, which may be easily cut and bent into different shapes. Black tie wire comes in rolls. Ask for the soft wire. This means it will stretch a little bit when you're pulling and cutting wire. You have to cut hundreds of pieces of black wire, 7 inches to 8 inches long. The best way to keep track of the cut black wire is to put the pieces in plastic coffee cans.

Painting Sculptures

The kind of paint you need will be exterior latex. You do not need to purchase paint by the gallon. The quart size, and even smaller, is okay for these projects. Just remember, if you get white and black, you can mix the two together in varying amounts and get a lot of different shades of grey. The same goes for other colors; for example, red with white gives you pink, etc. When mixing different colors, I use plastic coffee cans. The old metal coffee cans turn rusty on the bottom. Have fun with the painting. I like having very bright colors on these statues. Brushes, of course, would be all different sizes from small artist brushes to 1" and 2" paintbrushes.

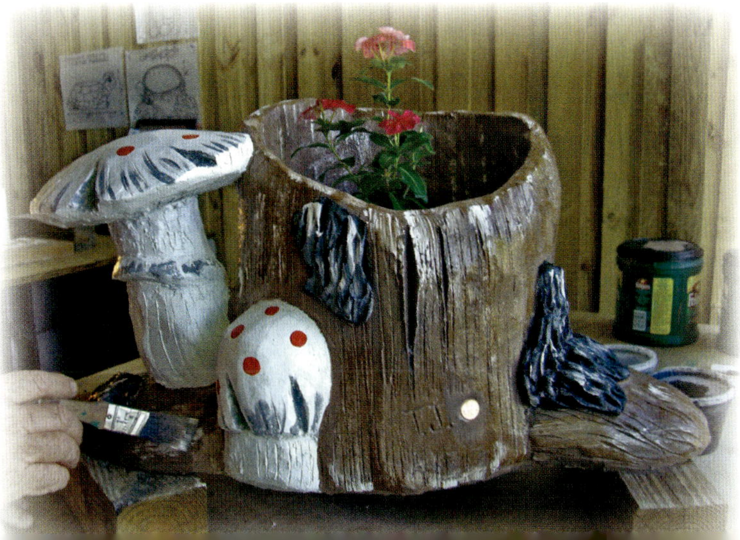

Turtle Planter

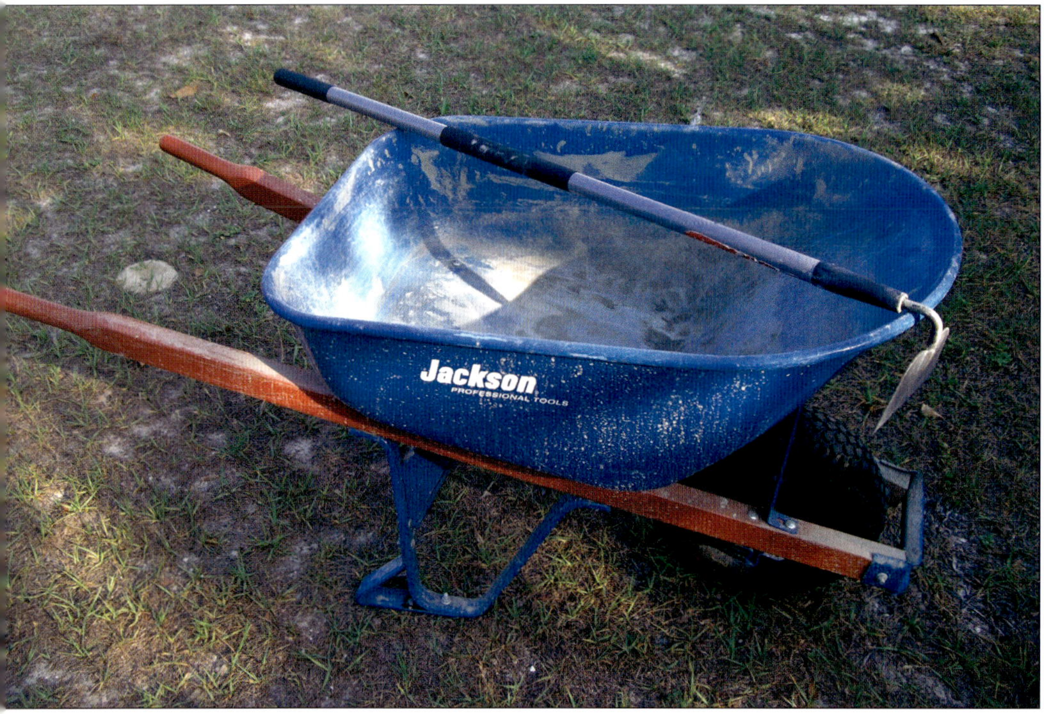

Wheel barrow and hoe for mixing mud.

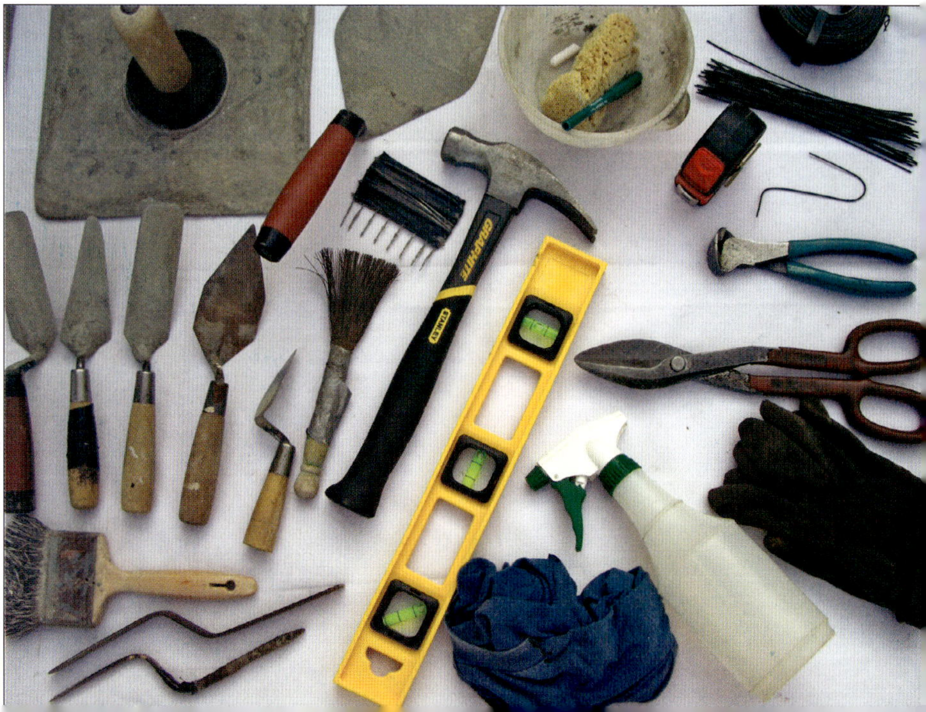

Wide variety of tools, including: gloves, shears, wire cutters, twenty-five foot tape measure, black tie wire, sponge, hammer, level, spray bottles, metal cement joints, different sized paint brushes, different sized point and margin trowels, metal scratcher, large mud trowel, hawk, and always have rags and plenty of Bandaids® on hand.

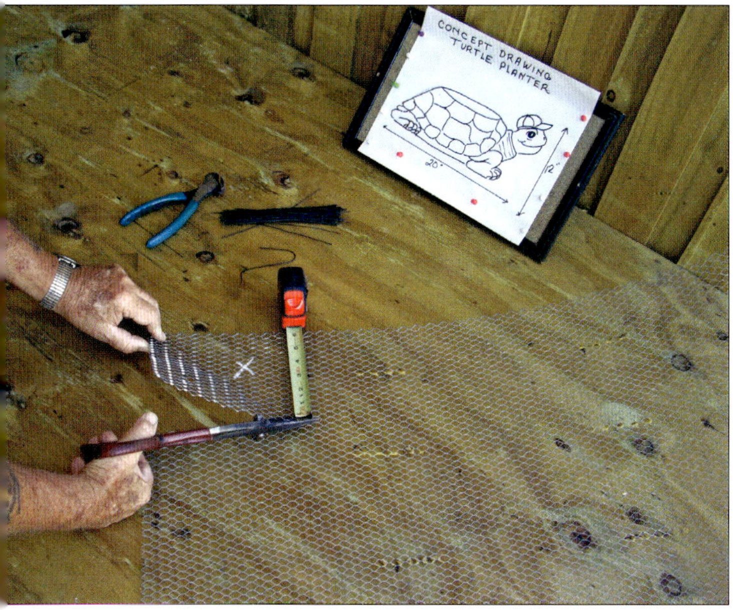

Beginning sculpture of turtle planter. Start by cutting a piece of lathe wire, approximately 6 inches wide by 36 inches long.

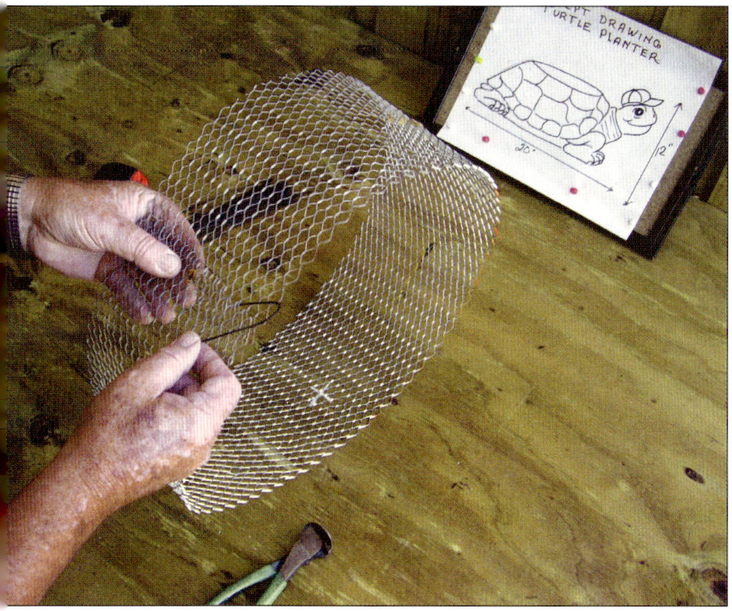

Roll the lathe wire, and overlap about 3 inches. Take a piece of tie wire and bend it into an S shape. Hold the bottom and push the top of the S through the lathe and bring it back through the lathe.

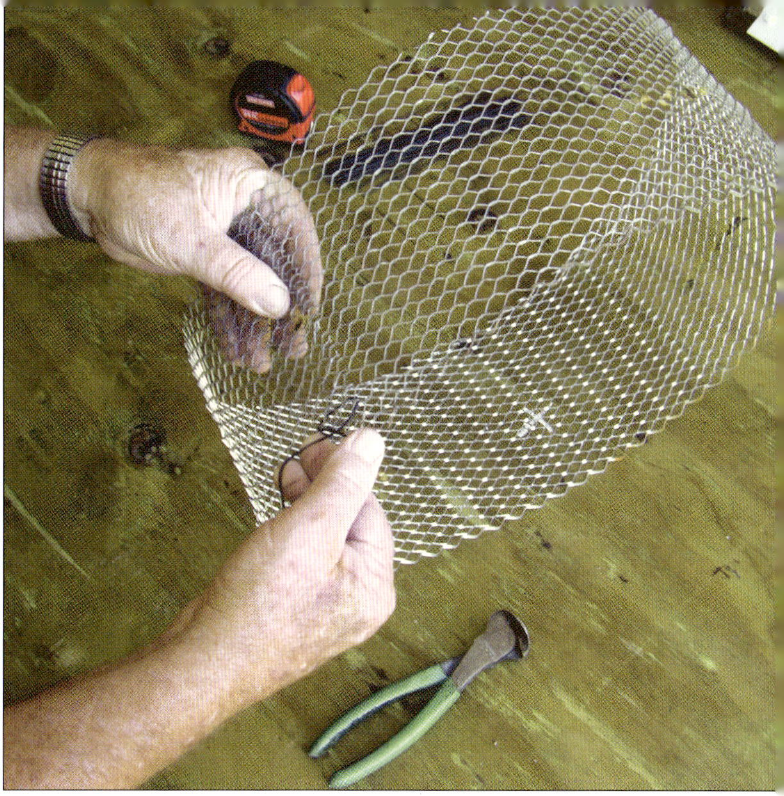

With your fingers, twist the wire once.

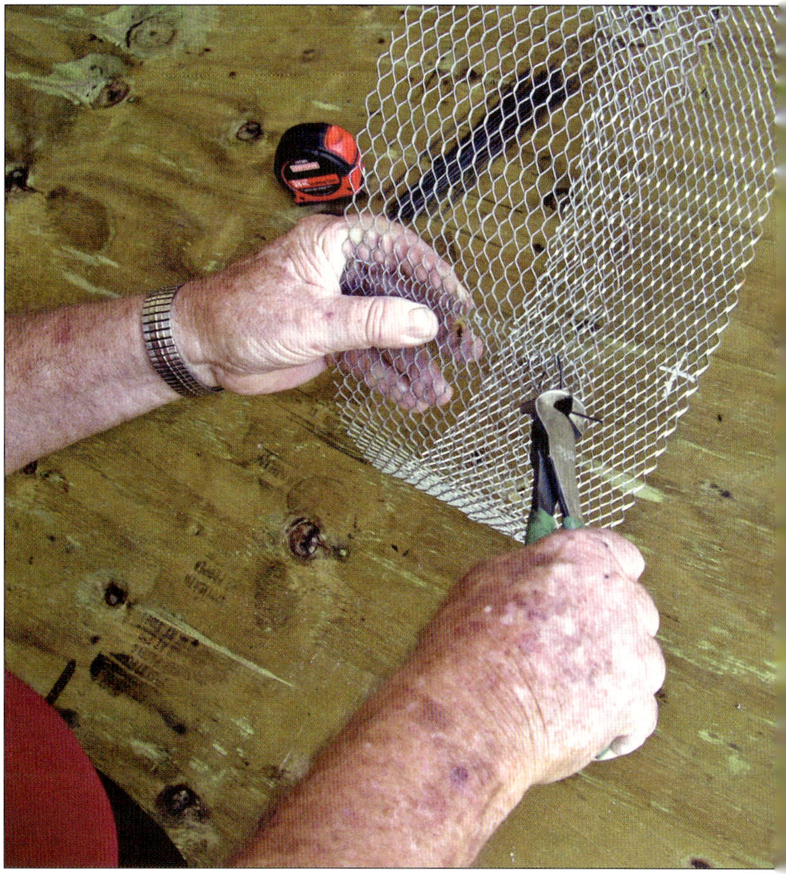

Using wire cutters, grab the twist and pull and turn two or three times and cut the excess off.

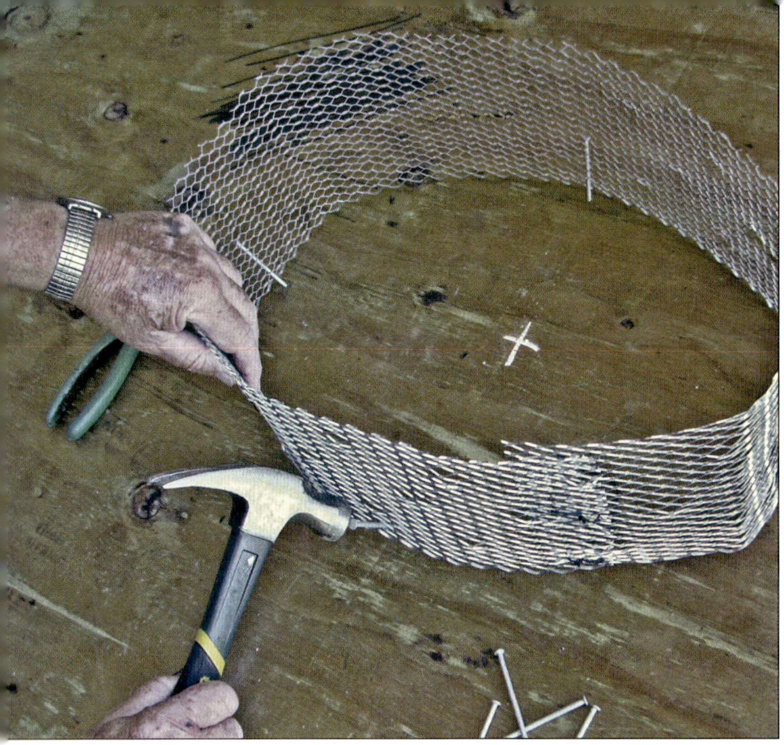

Take the circular shape and bend it into an oval shape. For this, you need a wooden table. Nail #10 penny nails at an angle around the wire. This will keep your sculpture stable while continuing to build your armature. If you do not have a wooden table that you can nail into, then use small concrete blocks or something heavy to hold the lathe in position.

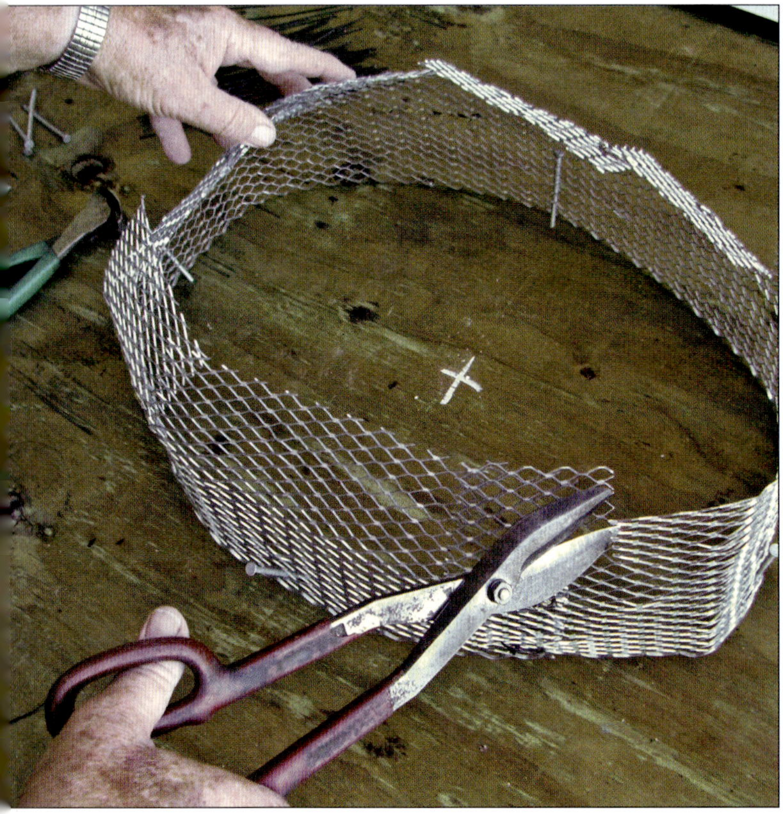

Cut around the top of the lathe 2 inches deep and bend the lathe in a little bit.

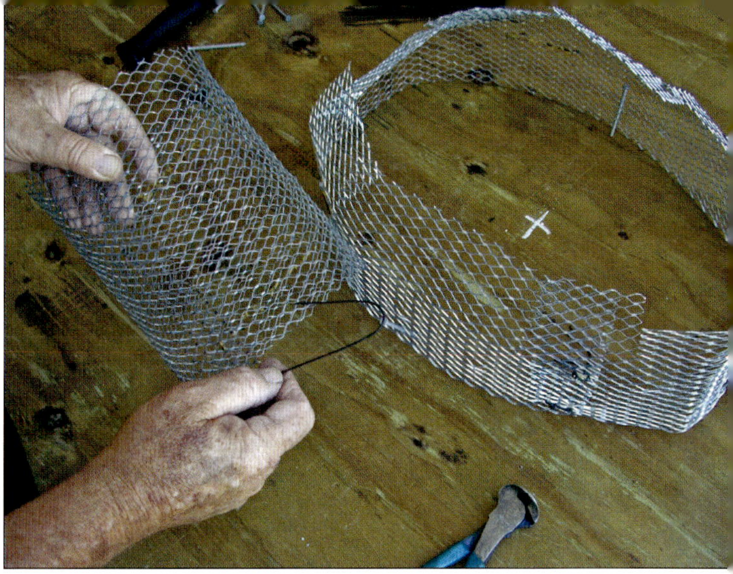

Cut another piece of lathe approximately 10 inches tall and 20 inches long. Roll it with a small overlap; tie with two or three pieces of tie wire. This piece is for the pot section.

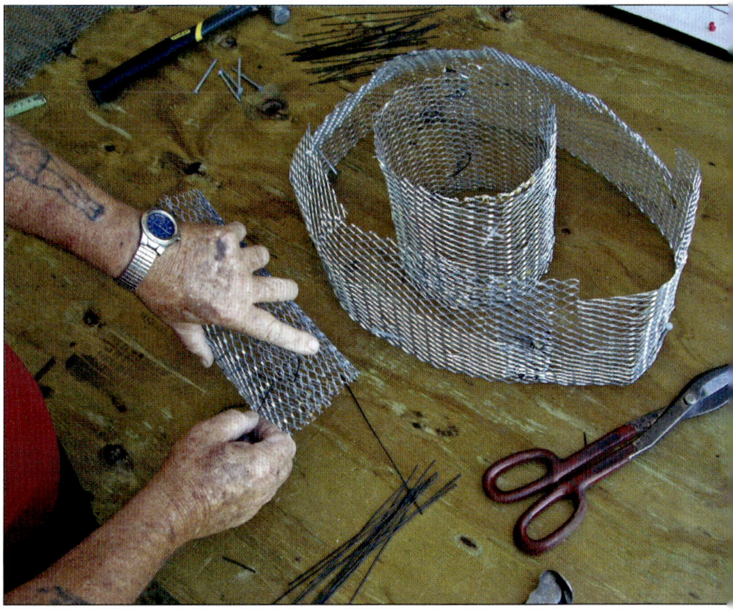

Cut another piece of lathe approximately 10 inches tall and 15 inches long. Roll into a small circle and tie it. This piece is for the neck. Put it aside for the moment.

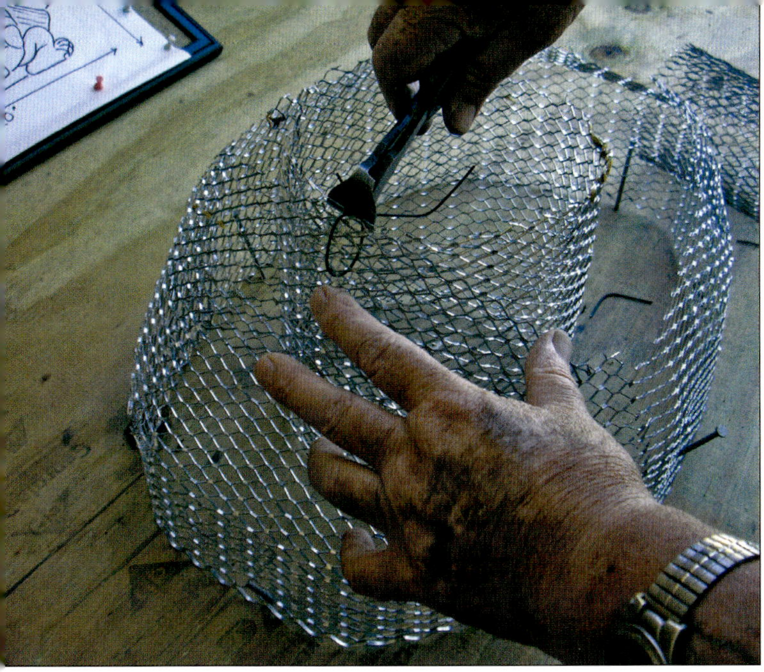

Start to tie the pot section to the front of the outer piece of lathe (oval) along the edges that were bent in earlier.

I'm cutting a small piece of lathe approximately 5 inches x 10 inches, rolling and tying together. At one end, close it up by cutting down 1 inch around the top and folding lathe down, securing with a tie. This is for the turtle's head. Put aside for now.

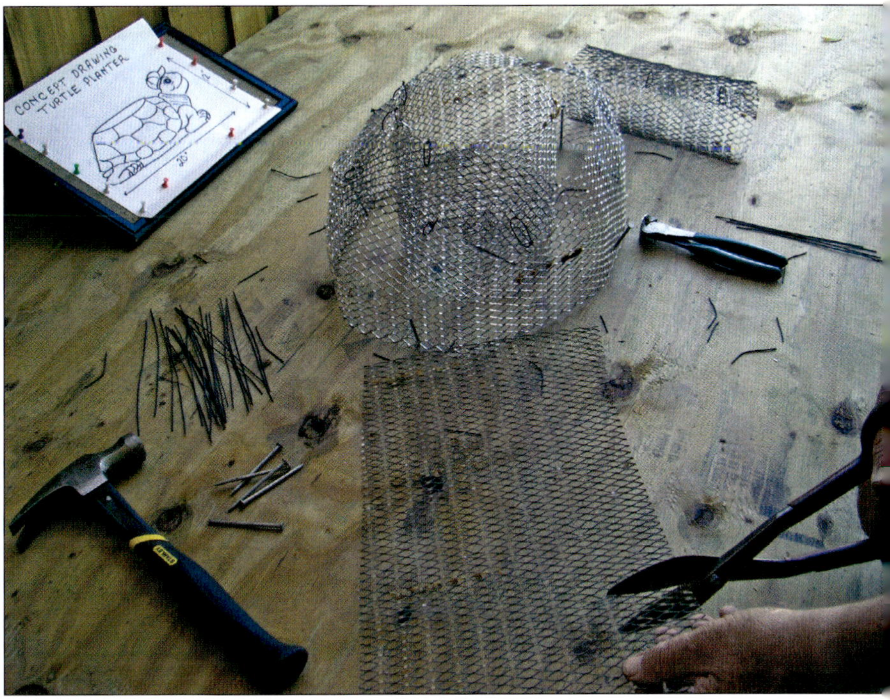

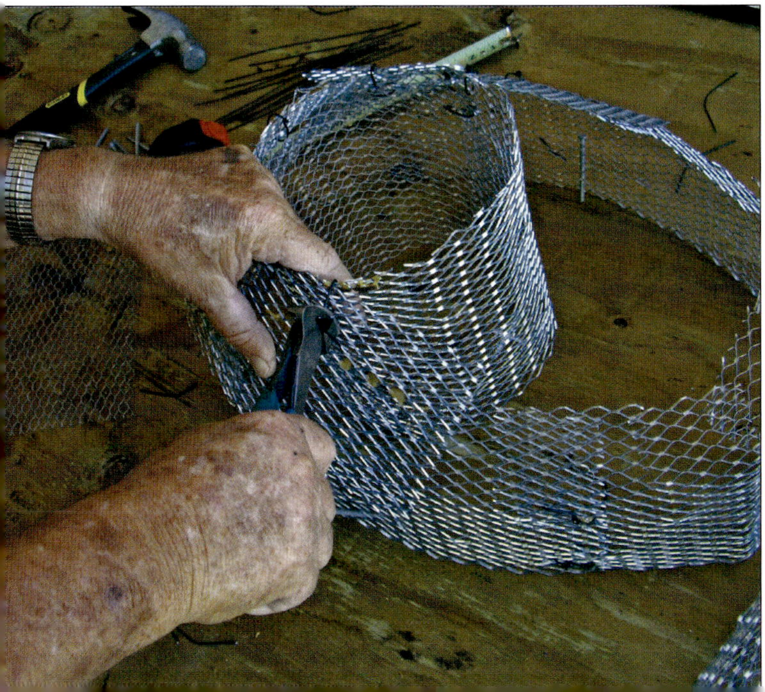

Continue to tie the pot section to the front of the outer piece of lathe (oval).

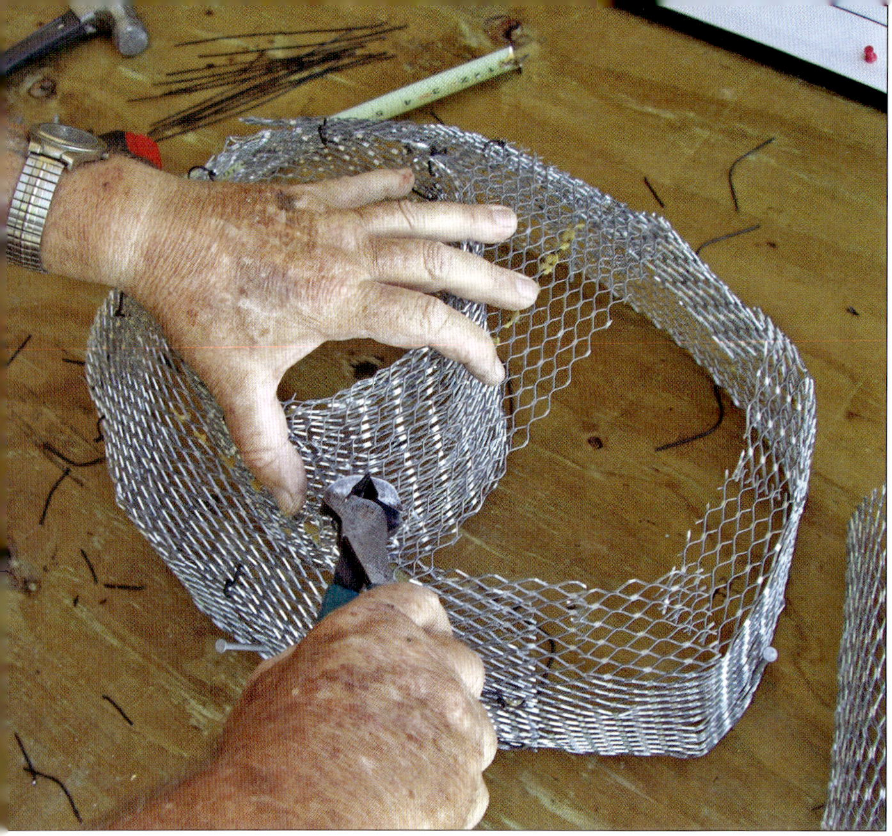

Continue to tie.

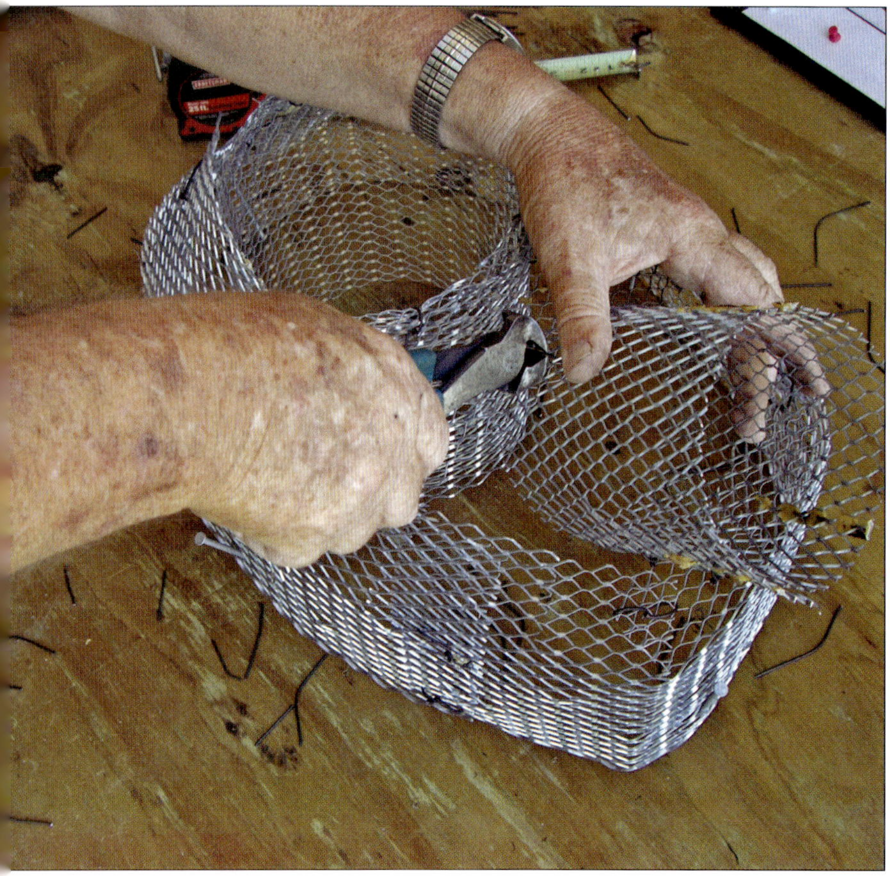

Take the piece of lathe that is for the neck. Put it in the open end of the oval. Angle one end up. Make a few ties.

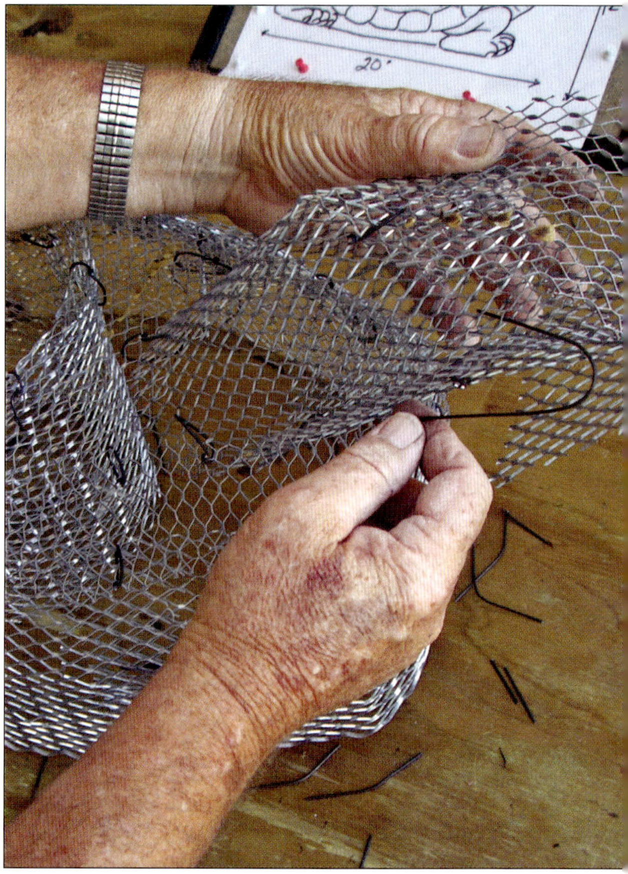

Take the piece of lathe that is for the head. Overlap it on the open end of the neck. Make a few ties.

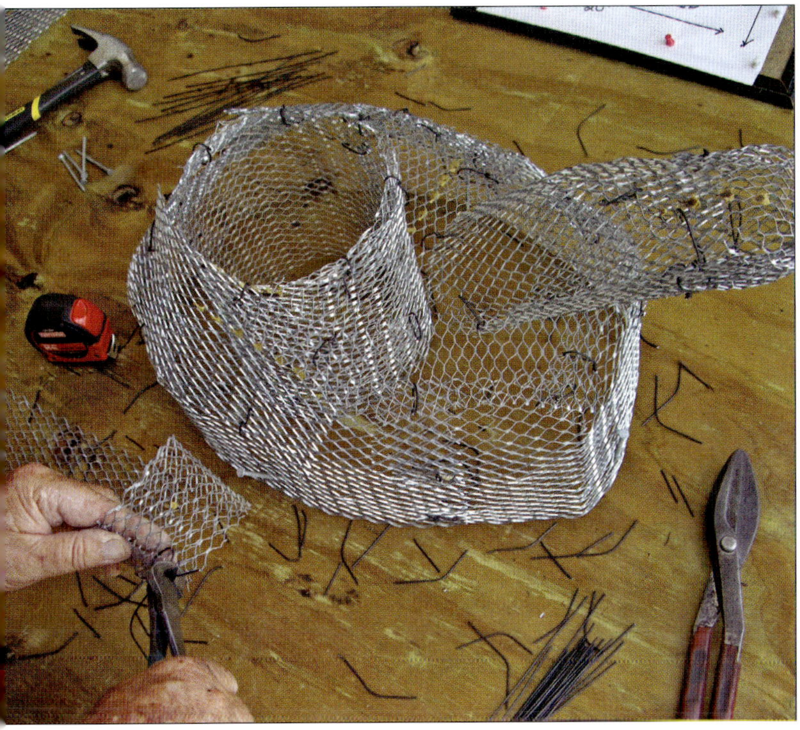

I am now tying a small piece of lathe for a baseball cap. You have to make your own judgment on the size of something like this. Just try to make it smaller than what you want it to look like with mud on it.

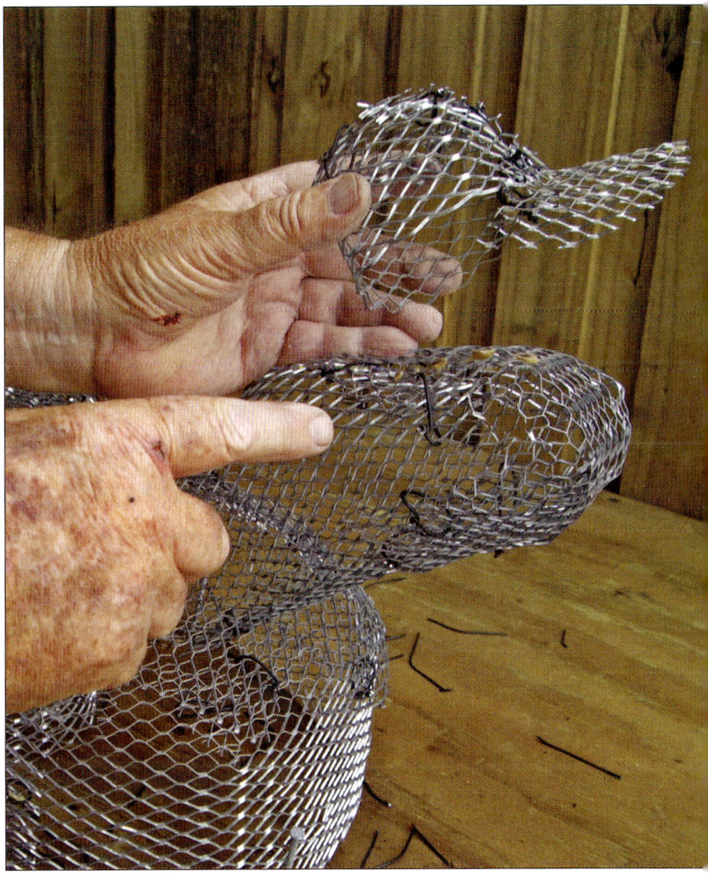

Place the hat on the head and secure with a couple of ties.

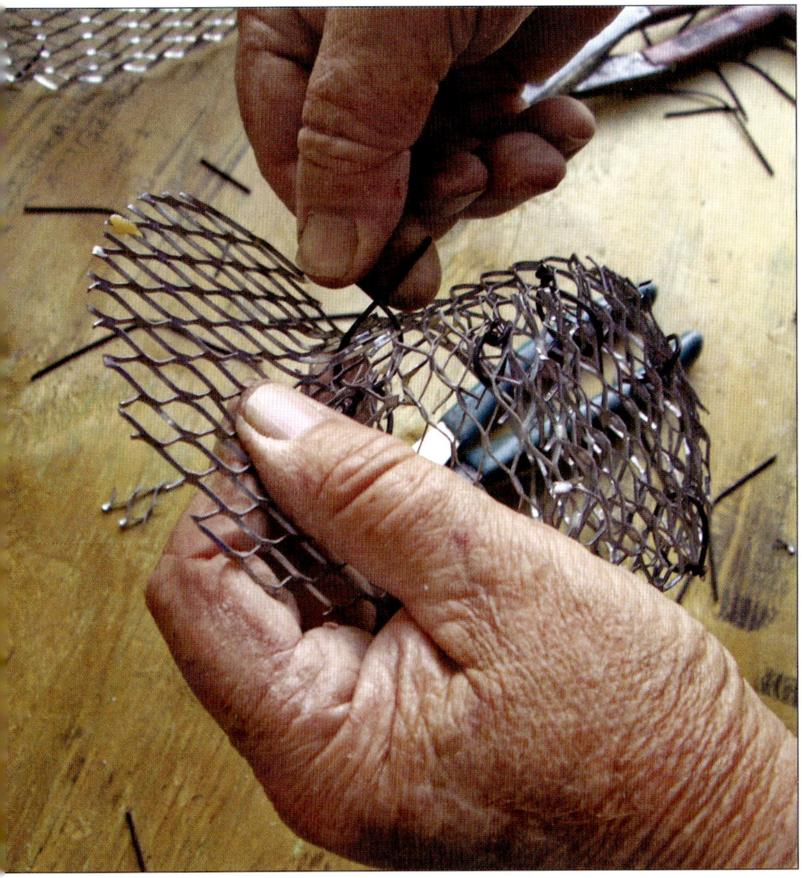

Tying the cap together.

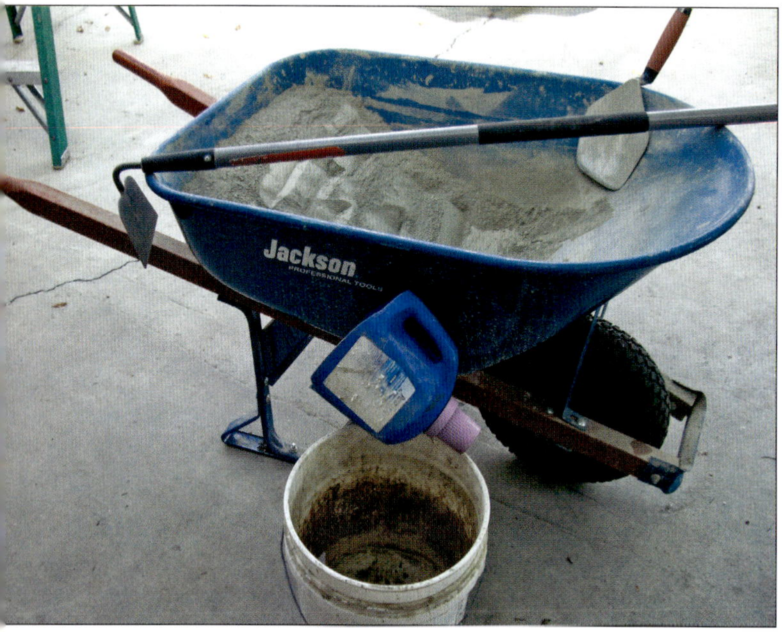

Wheelbarrow, hoe, mud trowel, water, and a half a bag of ready mix (Portland cement and sand). Pull ready mix to the high end of the wheelbarrow.

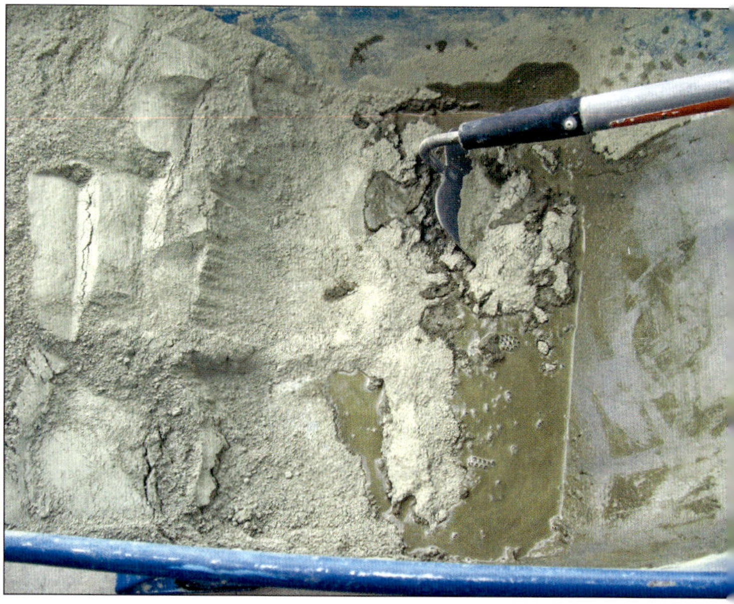

Using a hoe, pull the dry material down into the water, gradually mix together until you get a good consistency. Sometimes, the Portland cement sets up stiff; once or twice, you might have to spray a little water on it and re-mix.

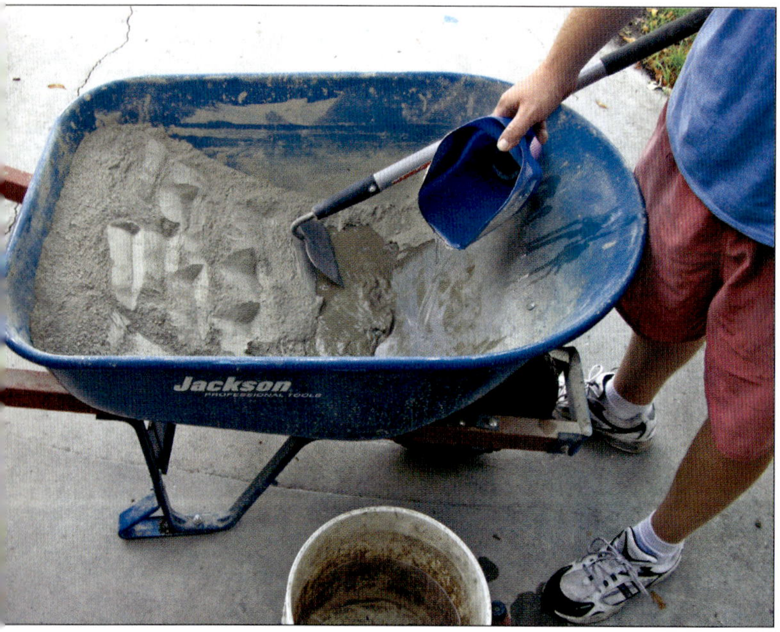

Using a liquid detergent bottle with the bottom cut off at an angle, scoop this container half full of water and pour the water into the low end of wheelbarrow.

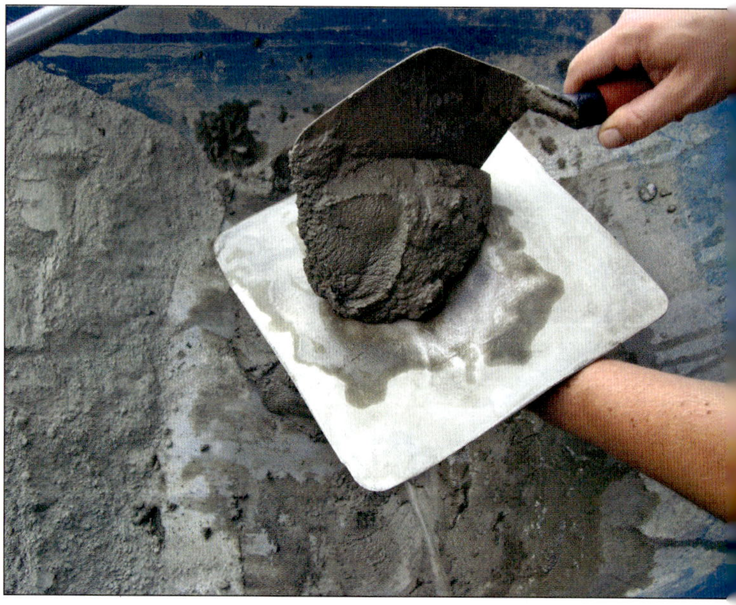

Using a mud trowel, place some mud on your hawk. You might want to use a small amount at first.

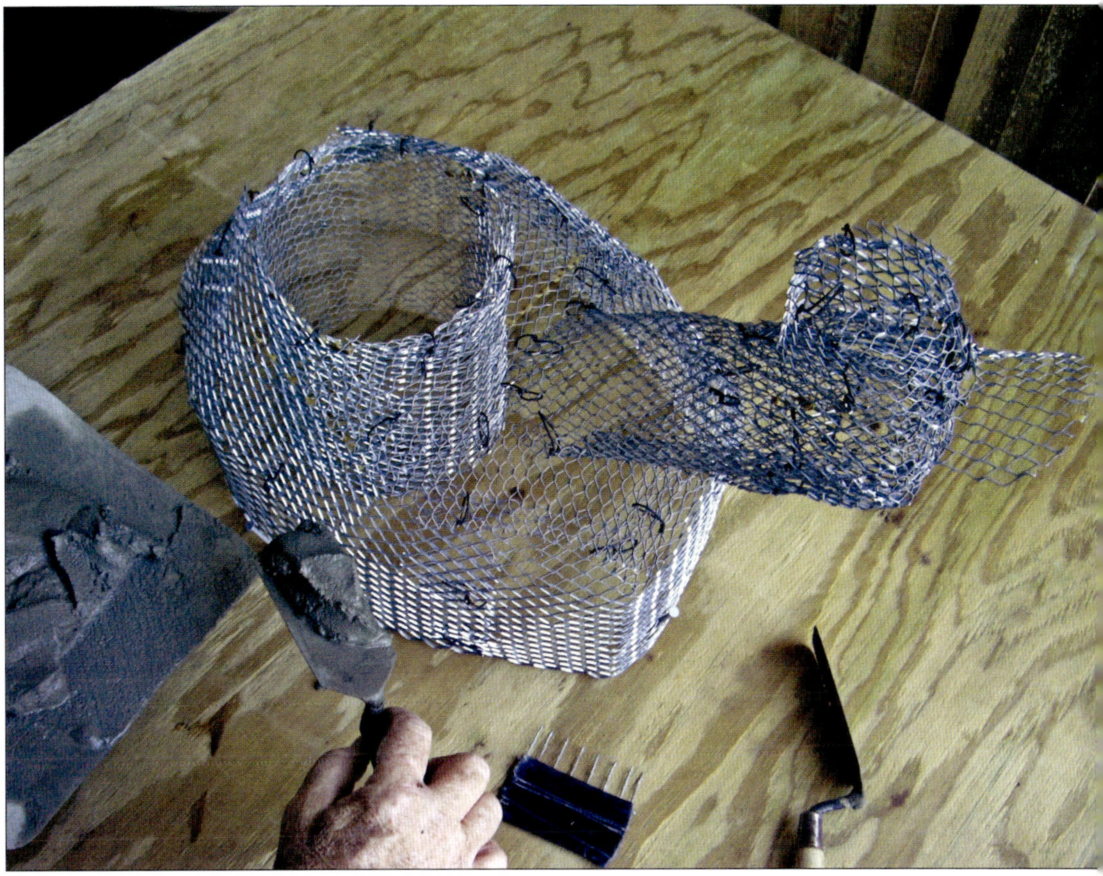

Using a small margin trowel, I'm picking up some mud.

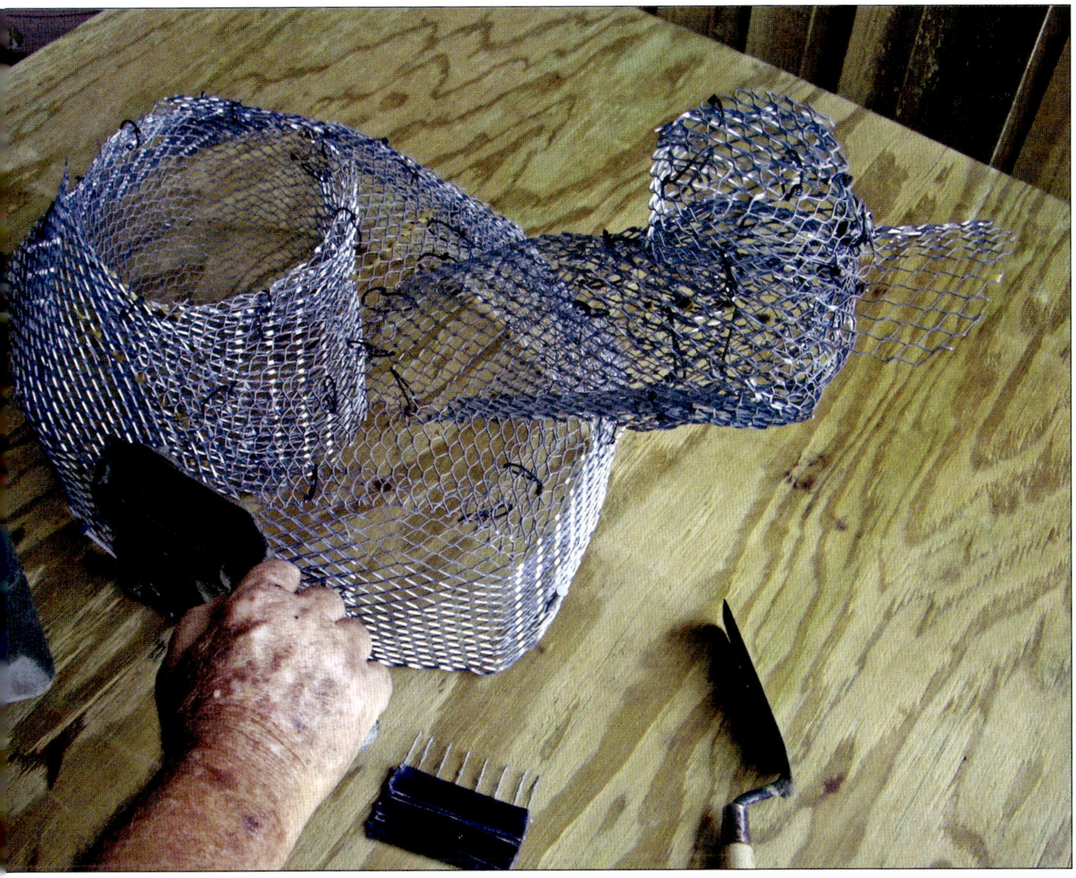

I'm pushing mud into the lathe around the bottom of the oval.

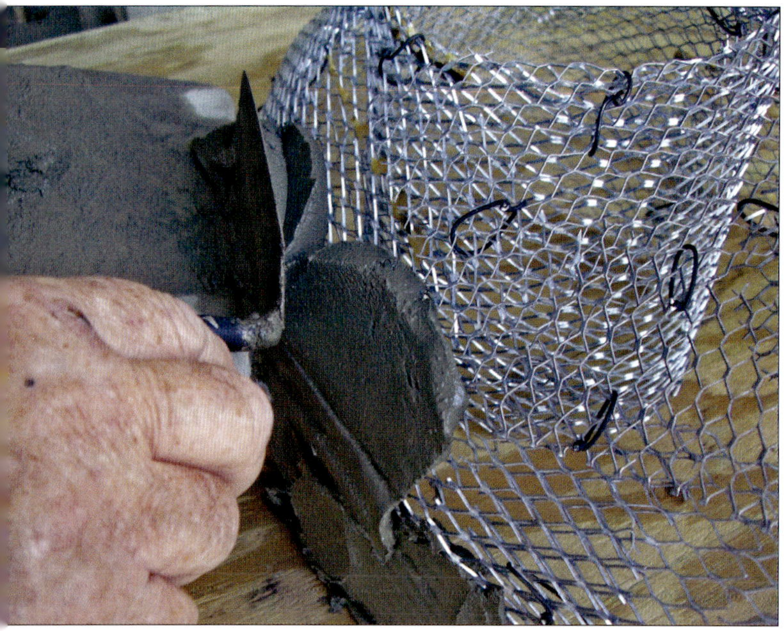

More mud.

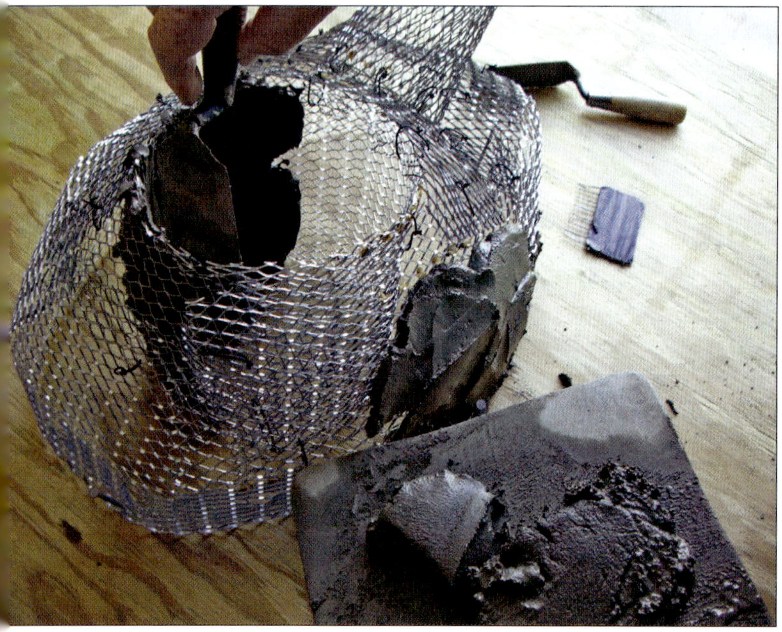

I am now placing mud inside the planter. You do not have to be very fancy here. You can pick up some mud in your hand and put it in the planter. In other words, as long as you get mud on the lathe, it can be rough (not to worry).

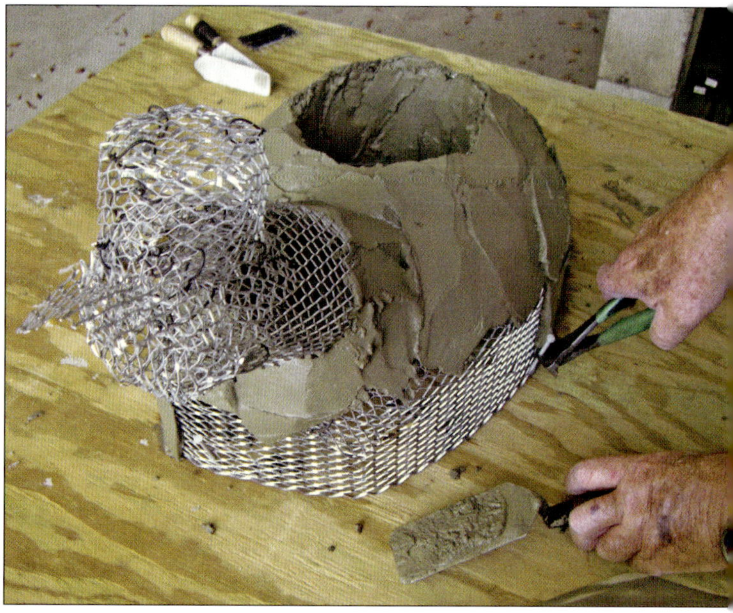

You can now take out nails by twisting and pulling up straight. The weight of the mud will stabilize the piece.

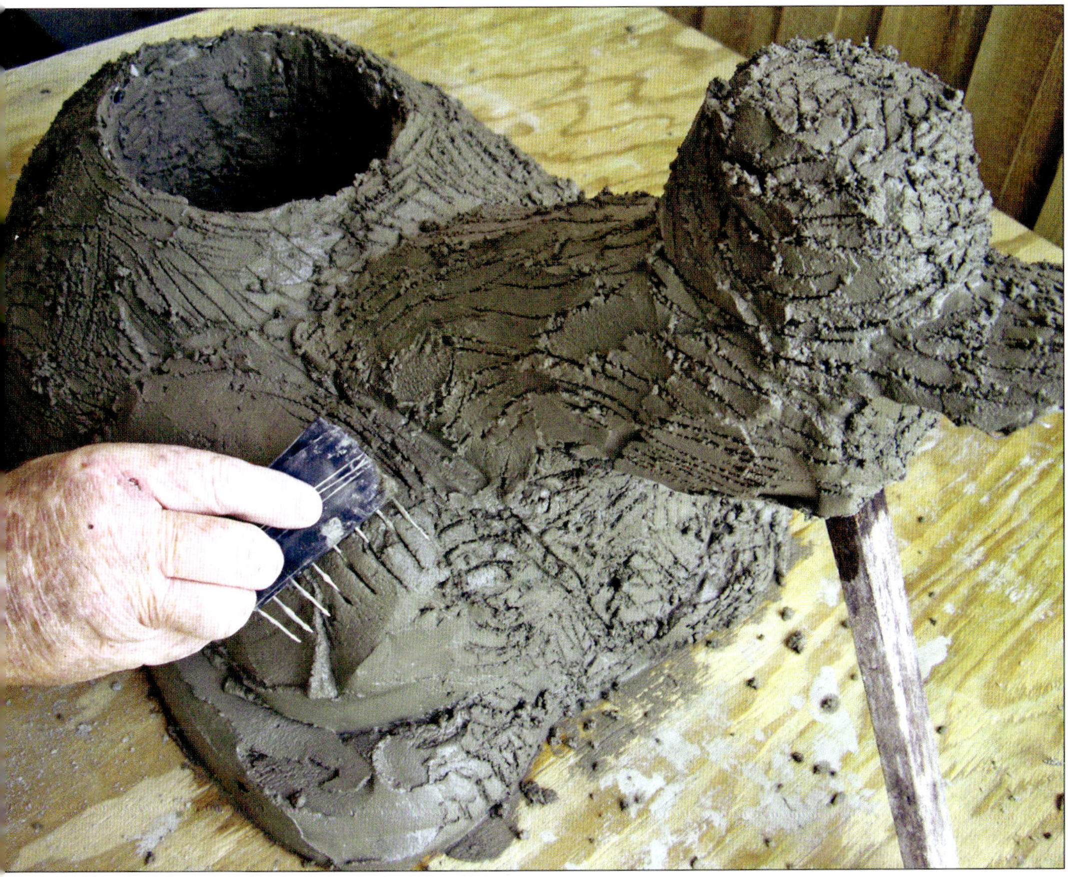

Before putting mud on the neck and head, take a small piece of wood (as shown) to support the weight of the mud while it is curing; then, you must scratch the whole sculpture with a piece of scratch wire.

You must remember that when you mix concrete, it's soft. In one day, it's hard. In two days it's five times as hard. Therefore, in two days, you can start your finish coat. Take out the supporting piece of wood.

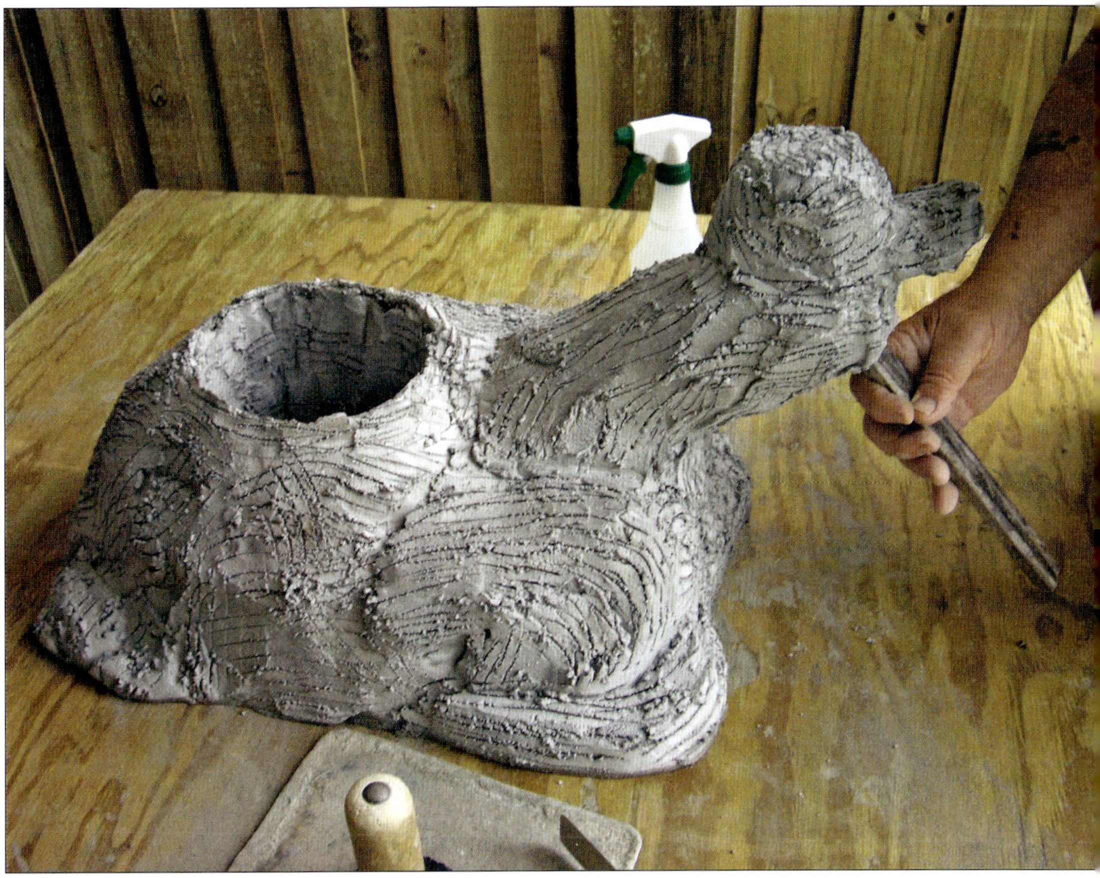

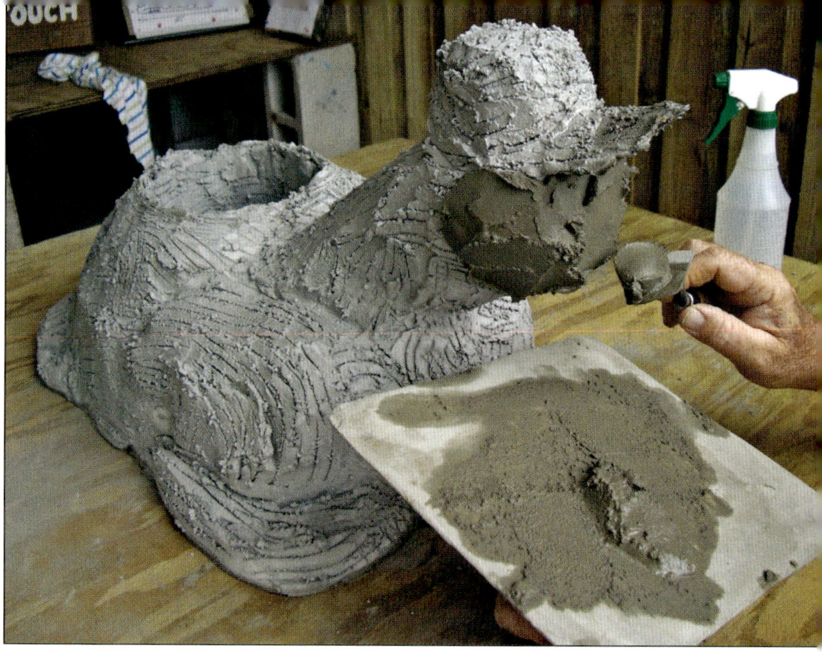

I start by adding mud to the face of the turtle. After I build it up a little bit, I'm going to leave it alone so it will set.

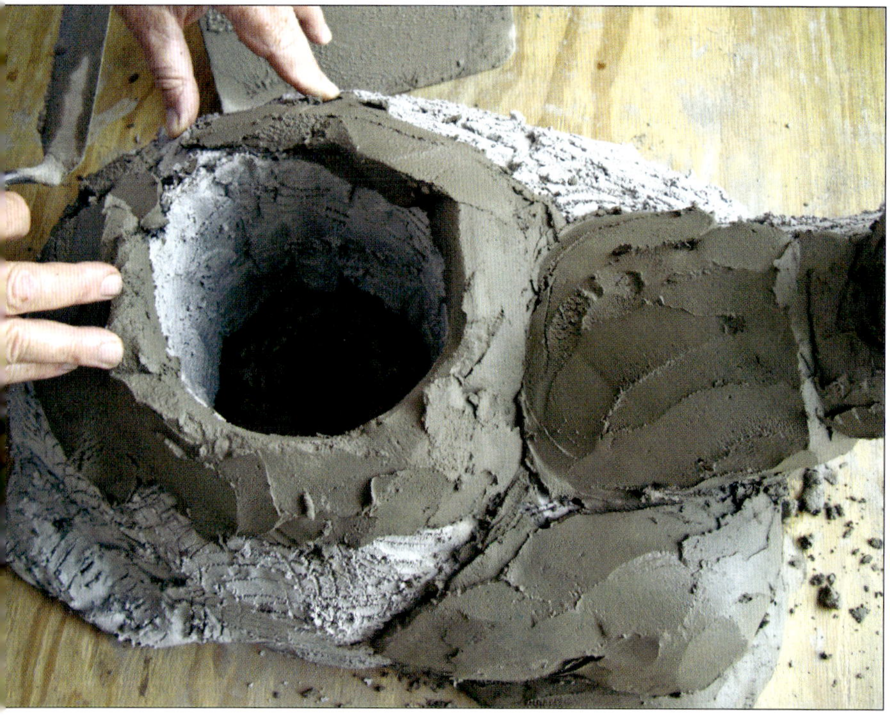

At this point, I am trying to level and make the top of the pot round.

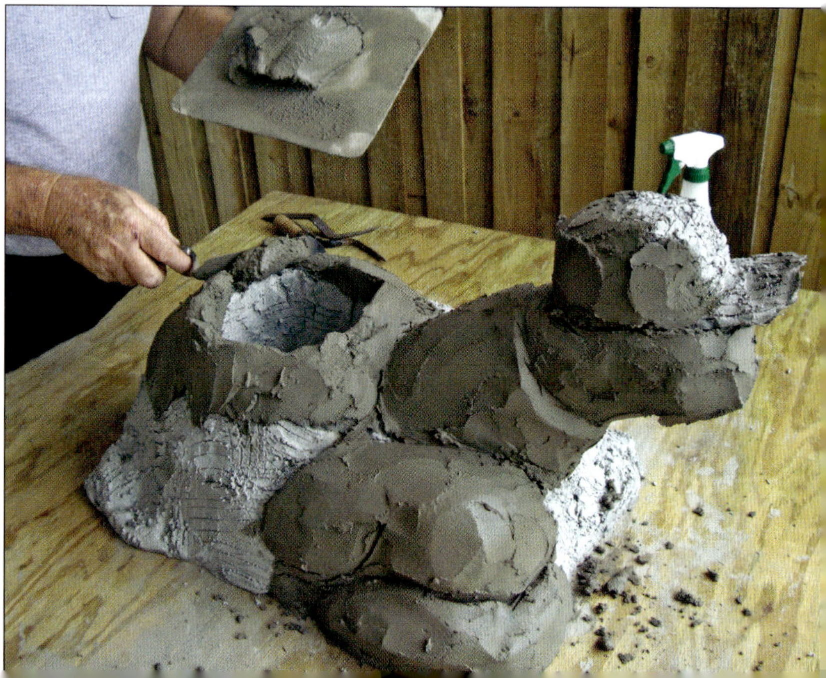

As you can see, I'm trying to make the edge of the baseball cap and the neck larger. Also, I am doing some work on the turtle's leg and foot; these can be built out in mud—sometimes using mud 2 inches to 3 inches thick. I always have a pile of old, bent, #10 penny nails and I just stick them into the mud for reinforcement. It works great.

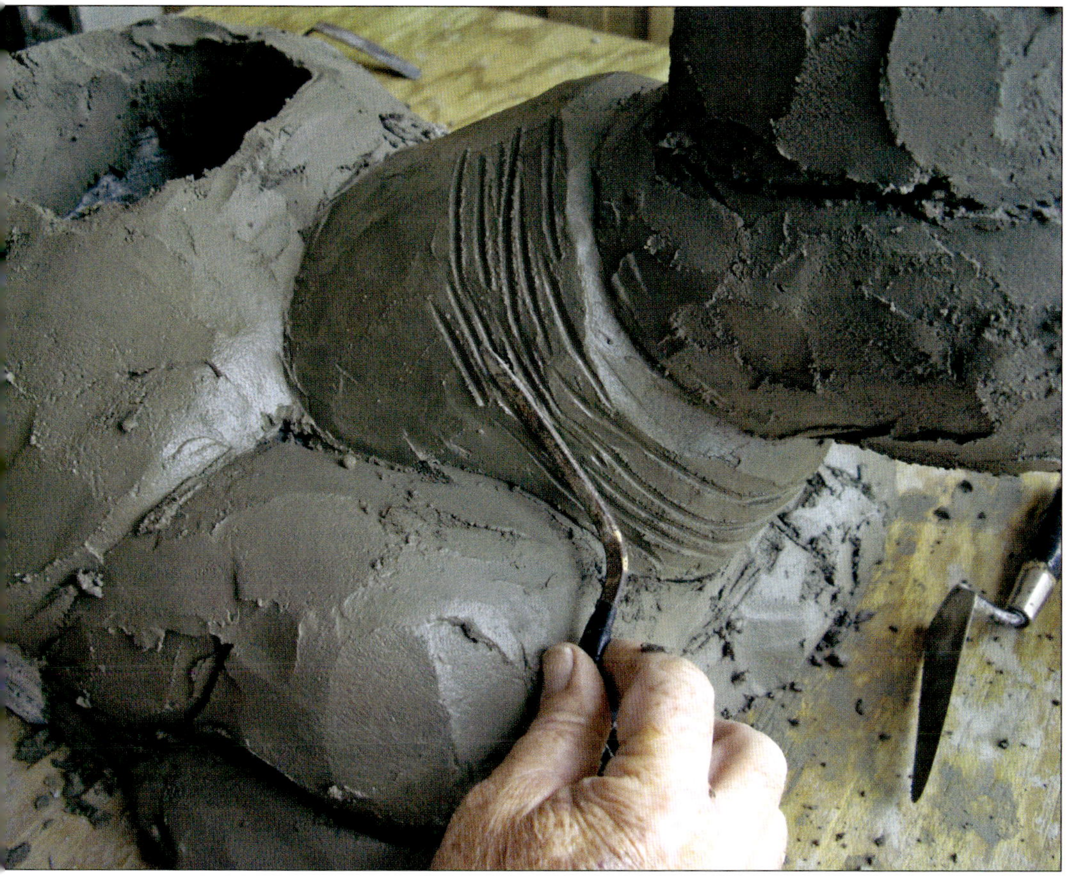

I'm using a small tool called a joint; masons use these tools when they are laying brick or blocks to smooth out mud in the joints. I have several different sizes that I use. You can buy these tools at masonry supply or building supply stores. I am putting grooves in the mud on the neck of the turtle.

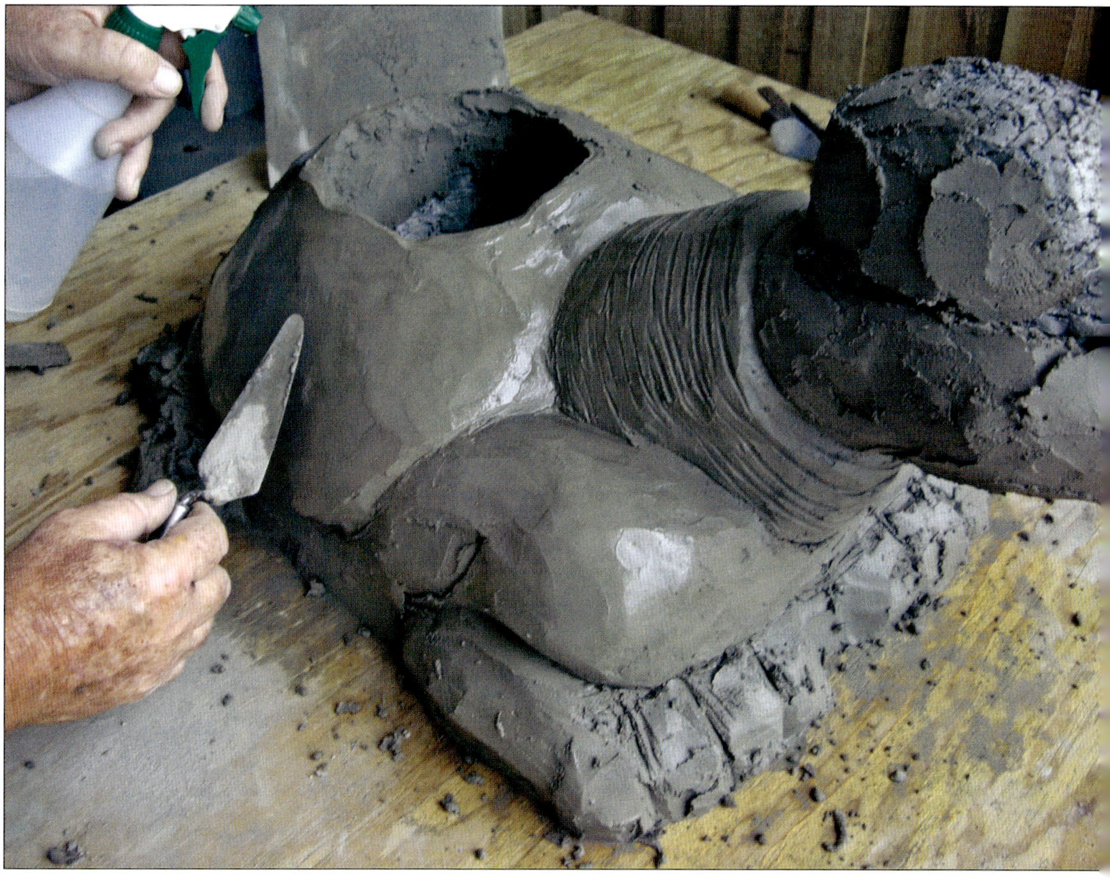

Adding more mud to the turtle shell.

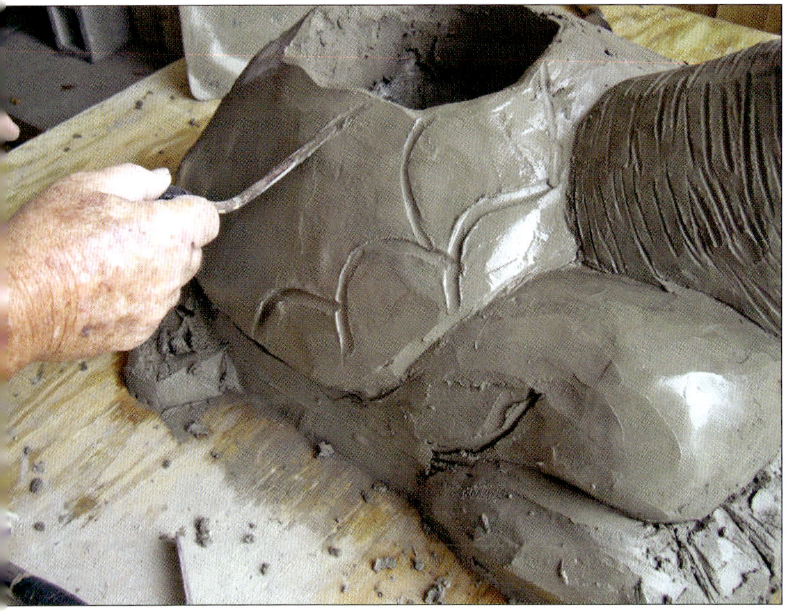

Making grooves in the turtle shell.

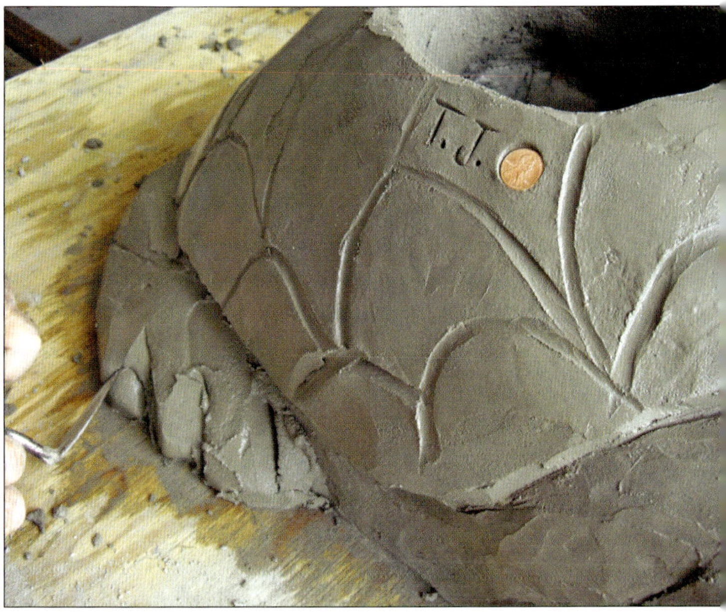

As an artist (sculptor), you must sign all your work and I hope you place a penny in the mud for luck. To keep the penny bright and shiny, spray it with clear lacquer ahead of time, otherwise they tarnish very quickly.

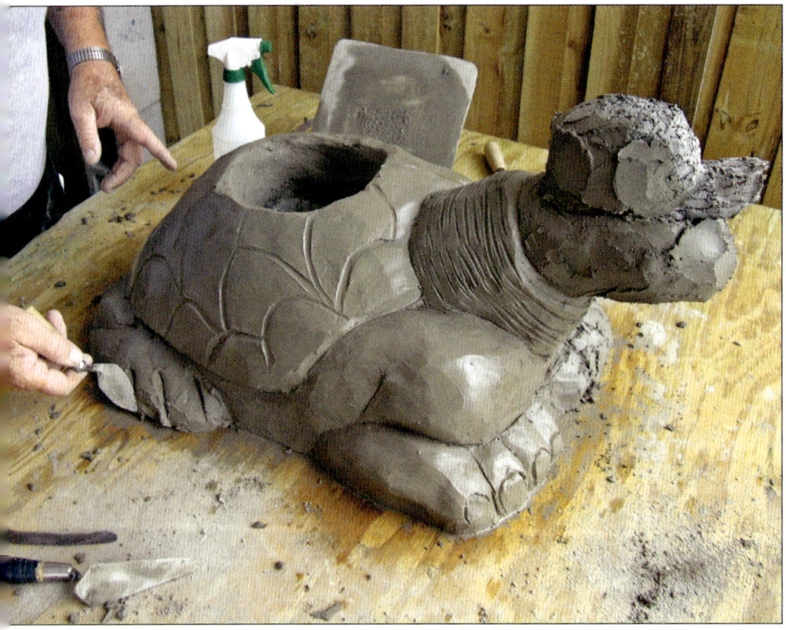

Making grooves for toes.

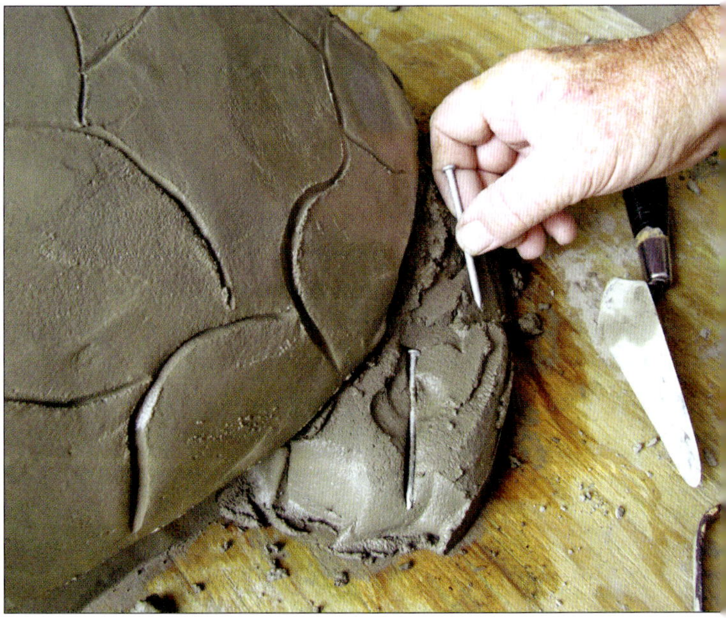

Placing galvanized nails, any size, as reinforcement in the foot of the turtle.

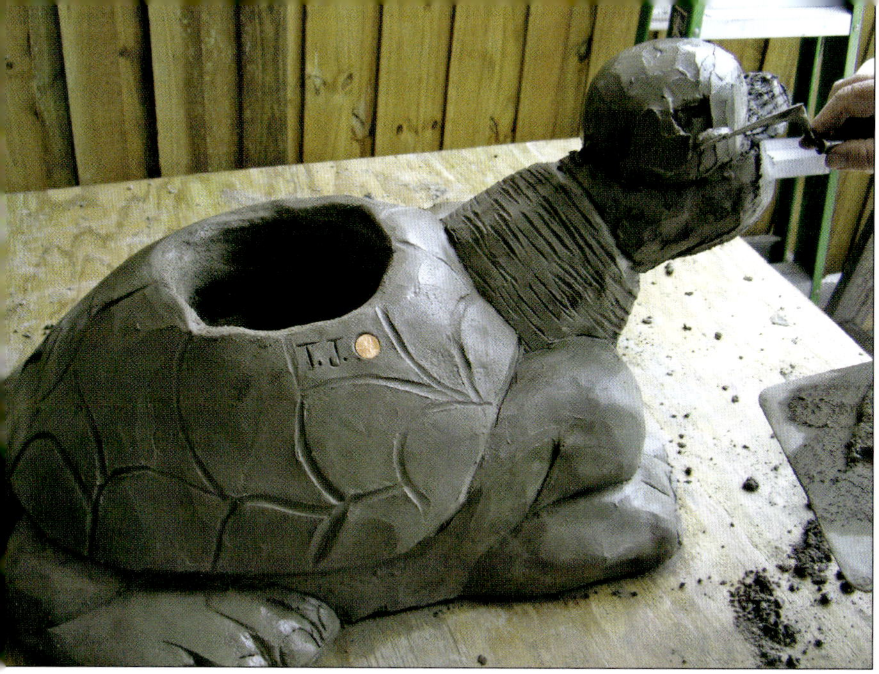

Working on the face and cap.

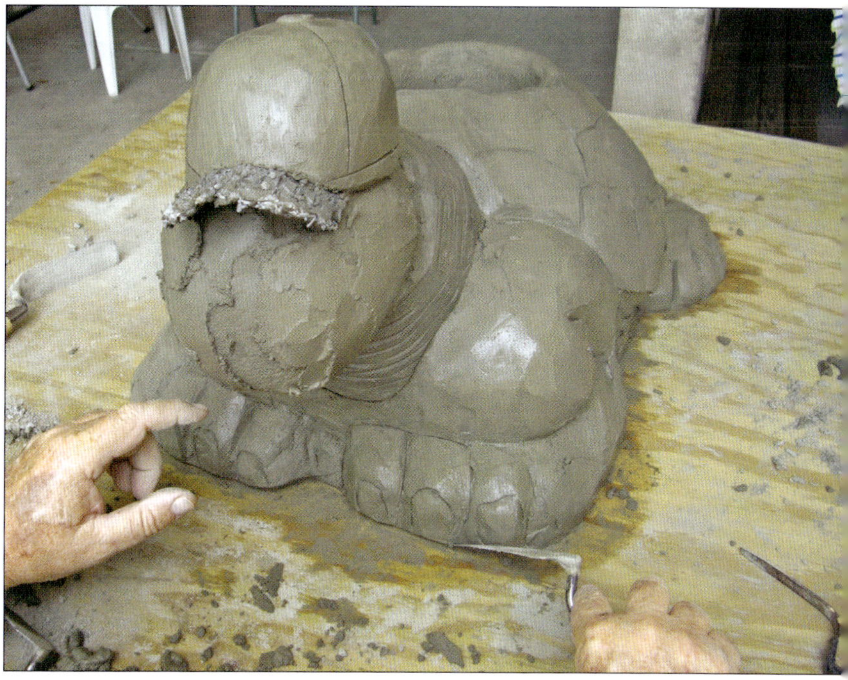

Working on the front toes.

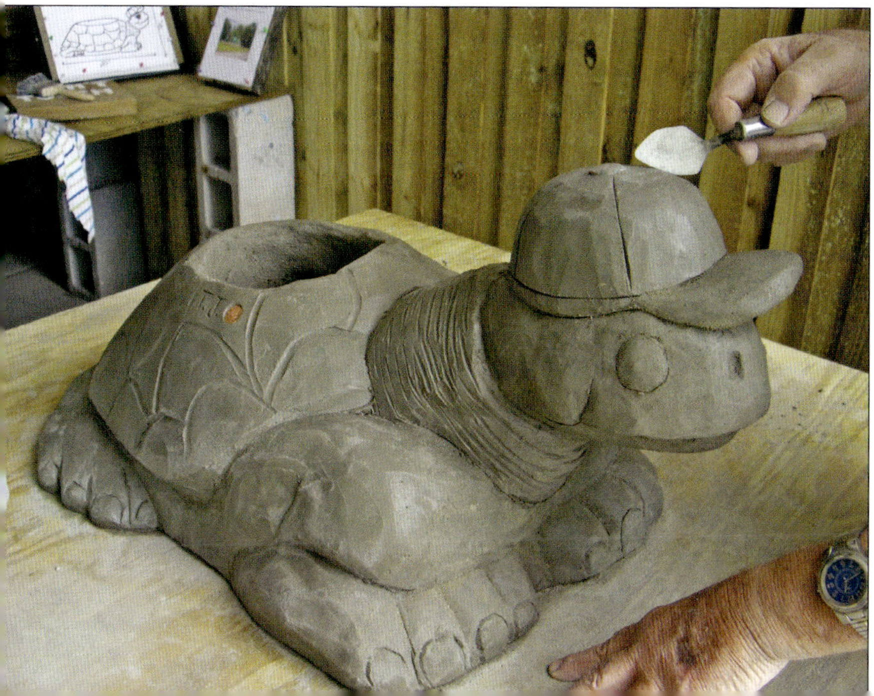

Putting grooves in the baseball cap. Mudding is complete.

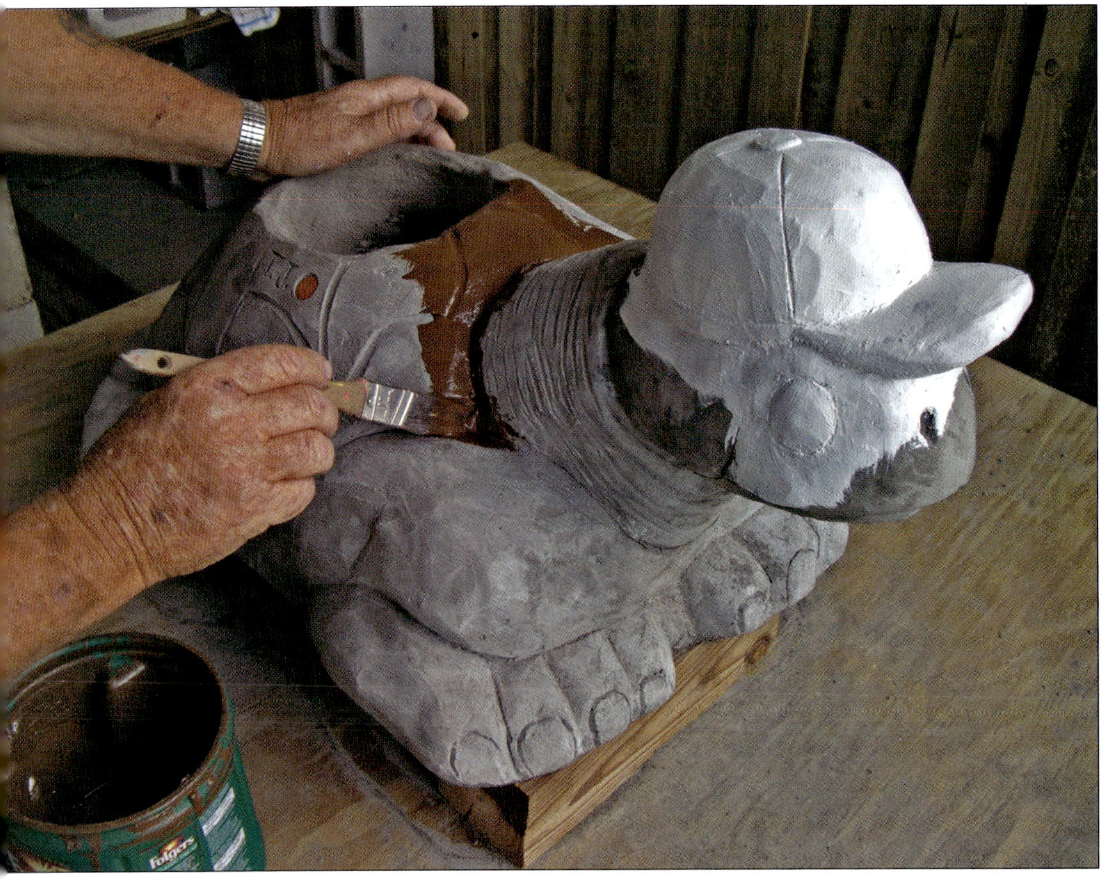

Let the concrete cure for six or seven days before painting. Now, have fun painting. Use plenty of bright colors.

Right side view of the completed sculpture (turtle planter).

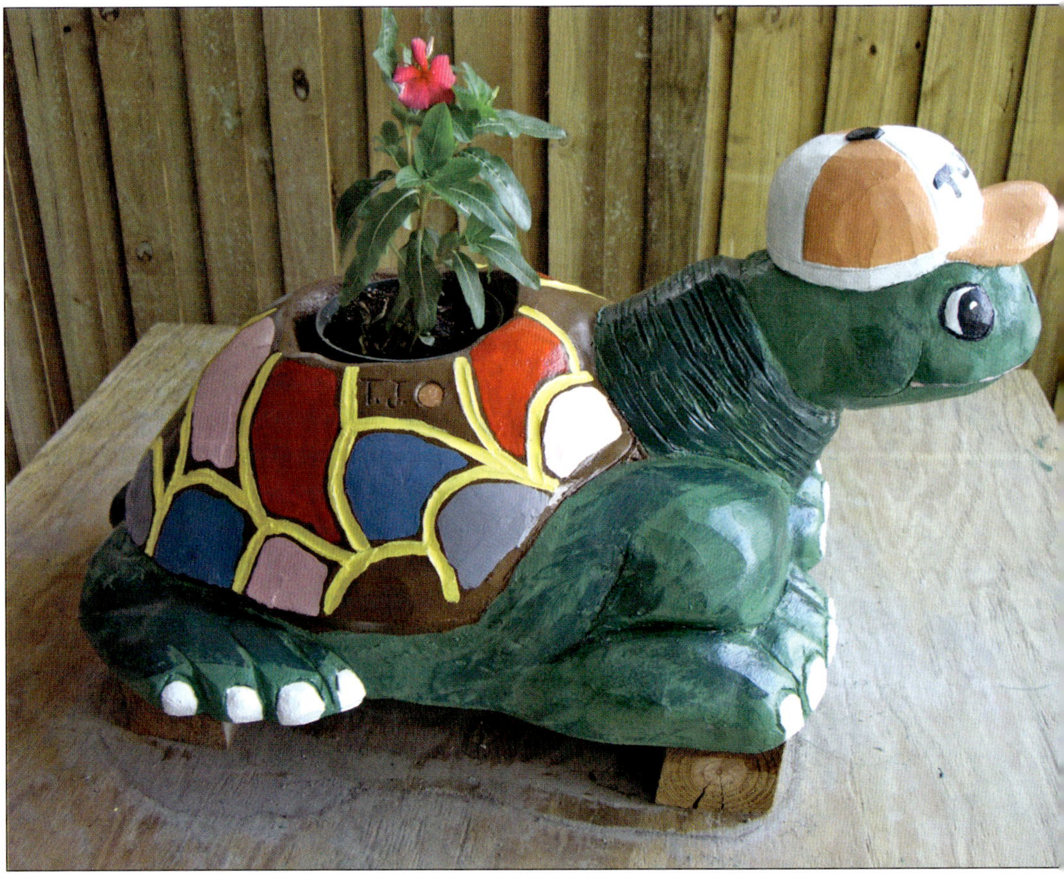

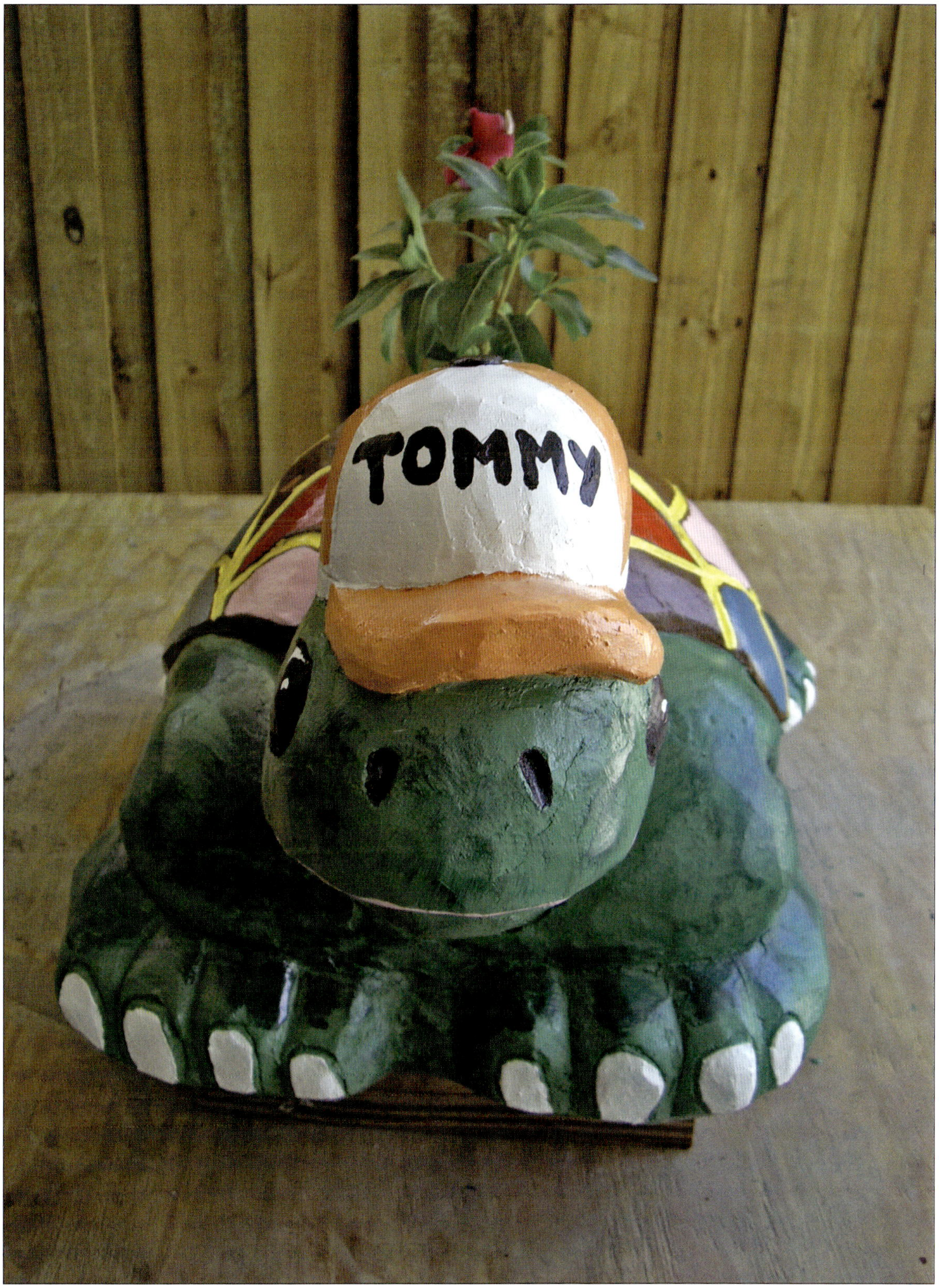

Front view of the completed sculpture (turtle planter).

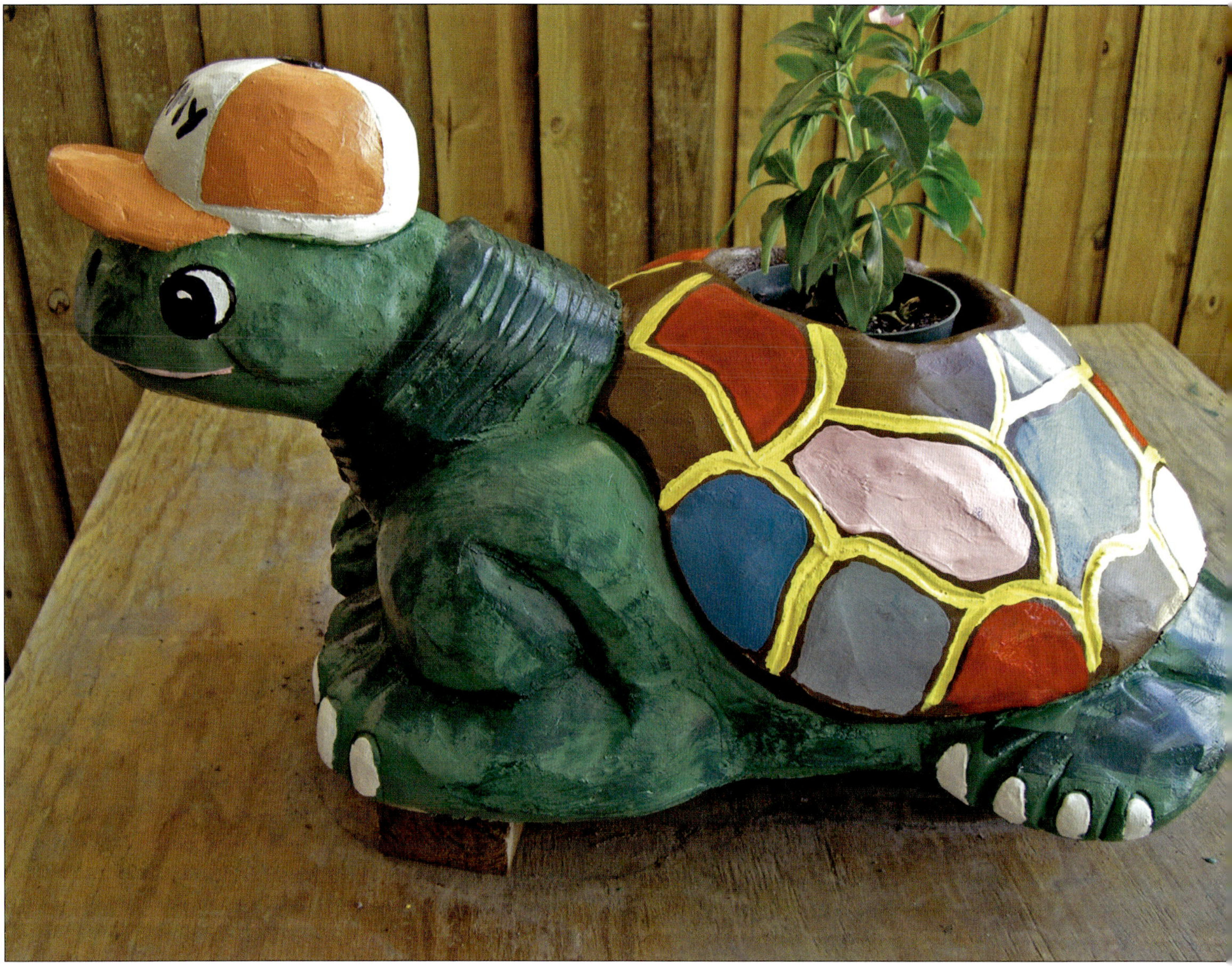

Left side view of the completed sculpture (turtle planter).

Gnome Planter

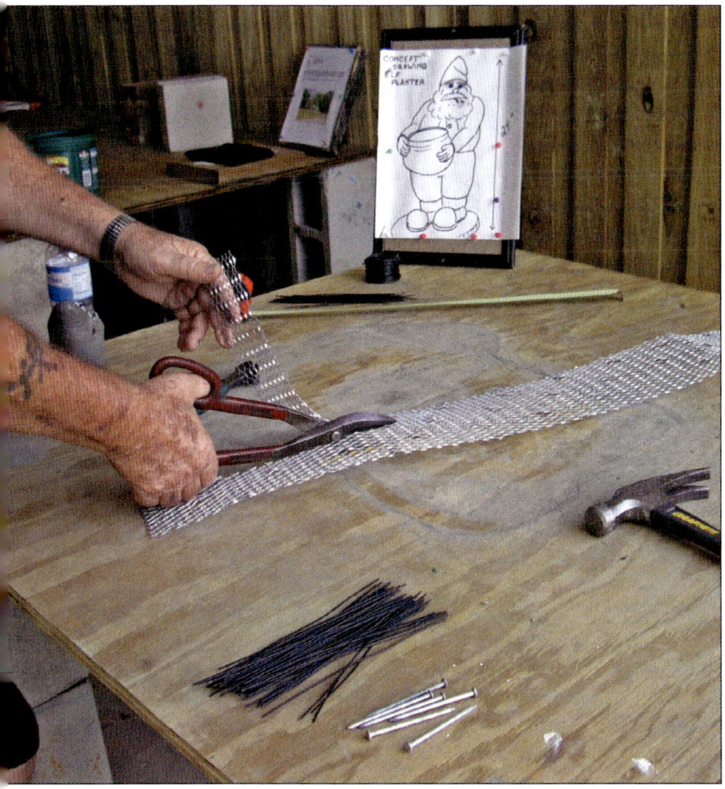

Cutting a piece of lathe for the base of the gnome, measuring 3 inches by 3'. Roll into a circle. Overlap and make a tie. Put aside for later. Now cut a piece of lathe 5 inches x 20 inches.

Roll lathe into a circle with a small overlap. Take a piece of tie wire, shape into an S. Hold bottom of S. Feed top through lathe.

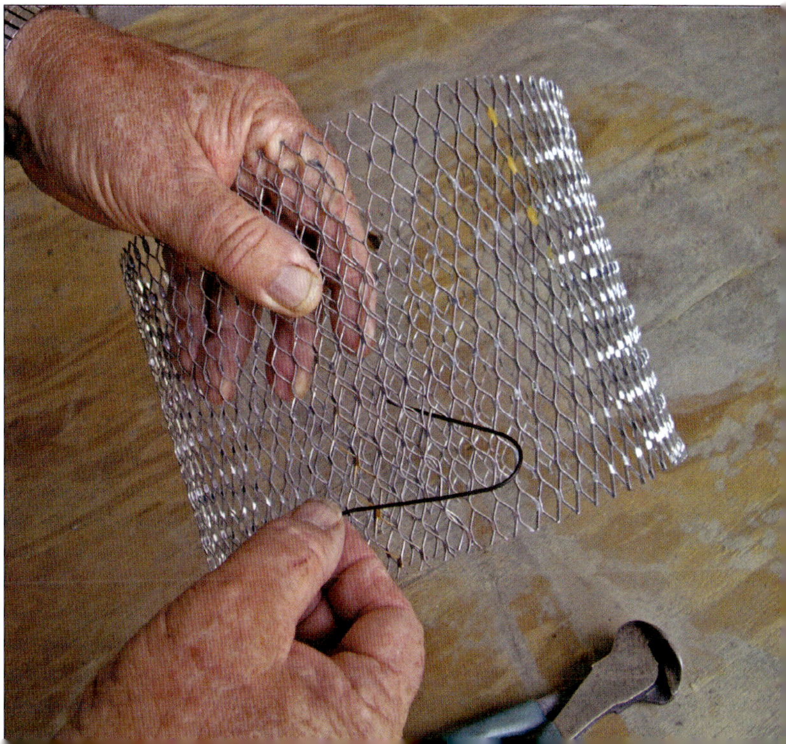

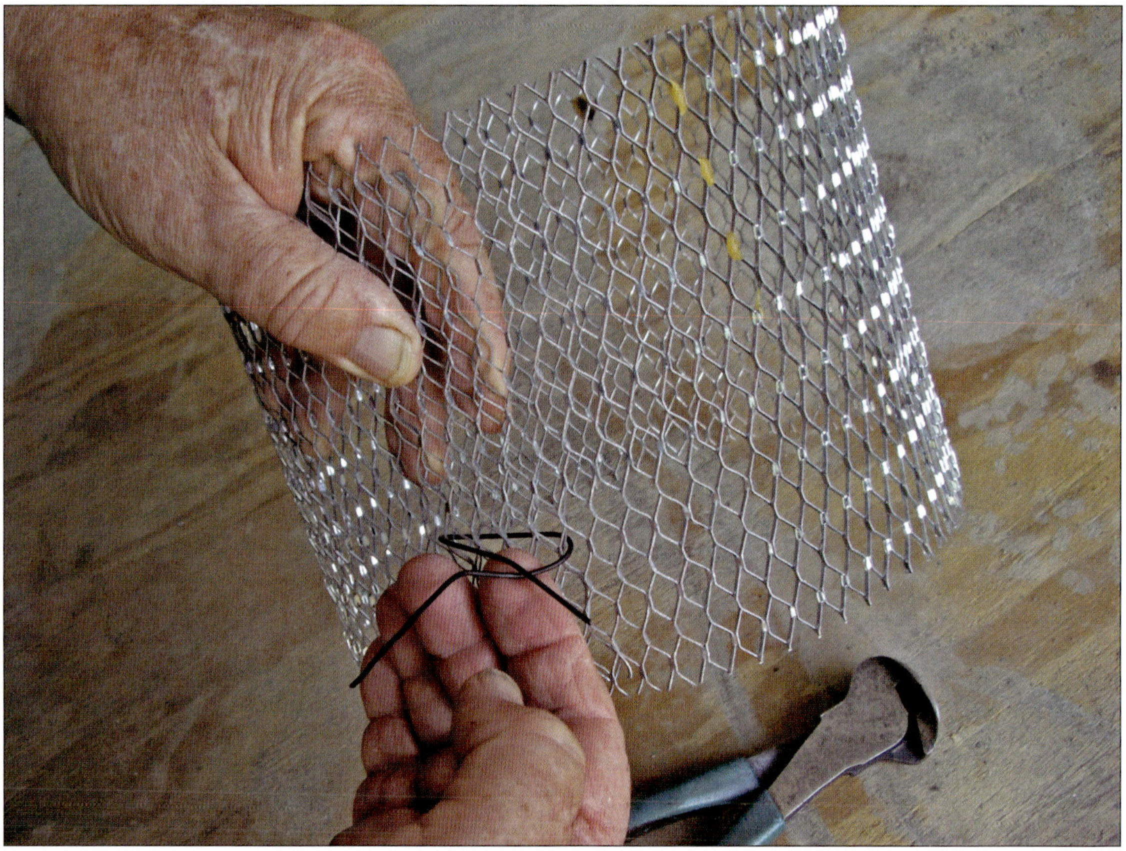

Pull wire through the lathe. Make a twist in the wire.

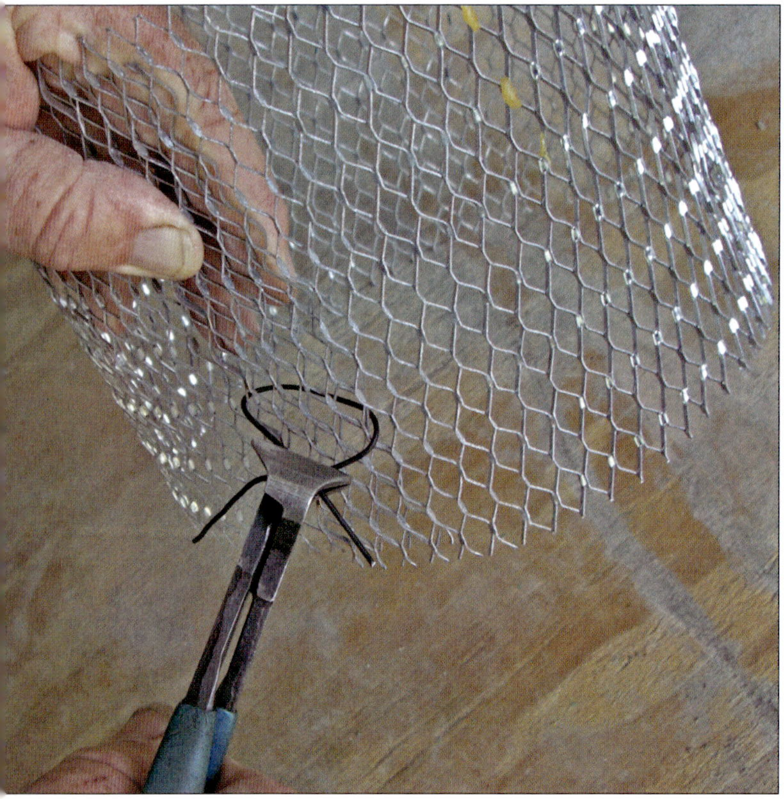

Grab twisted wire with wire cutters.

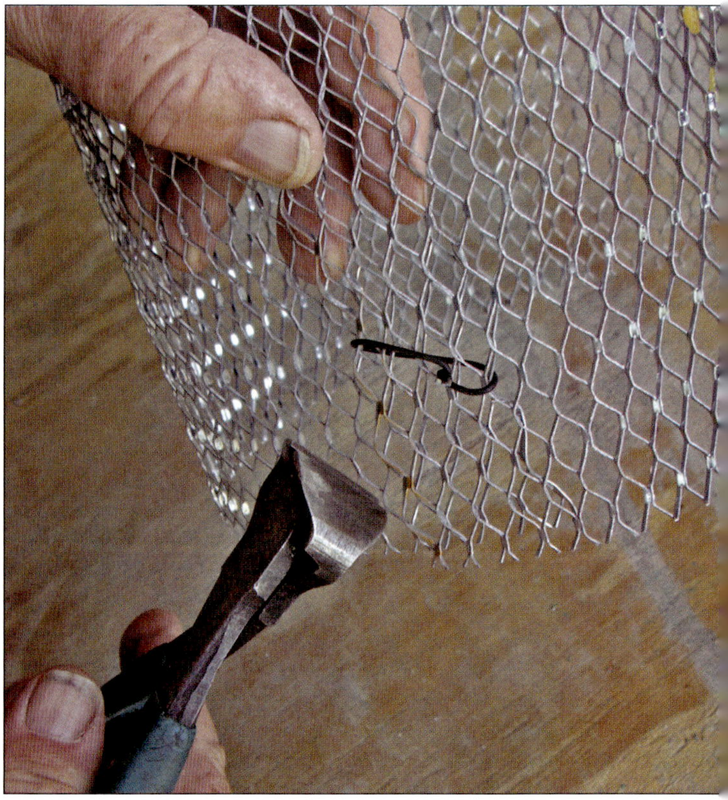

Turn two to three times to make it tight. Cut off excess.

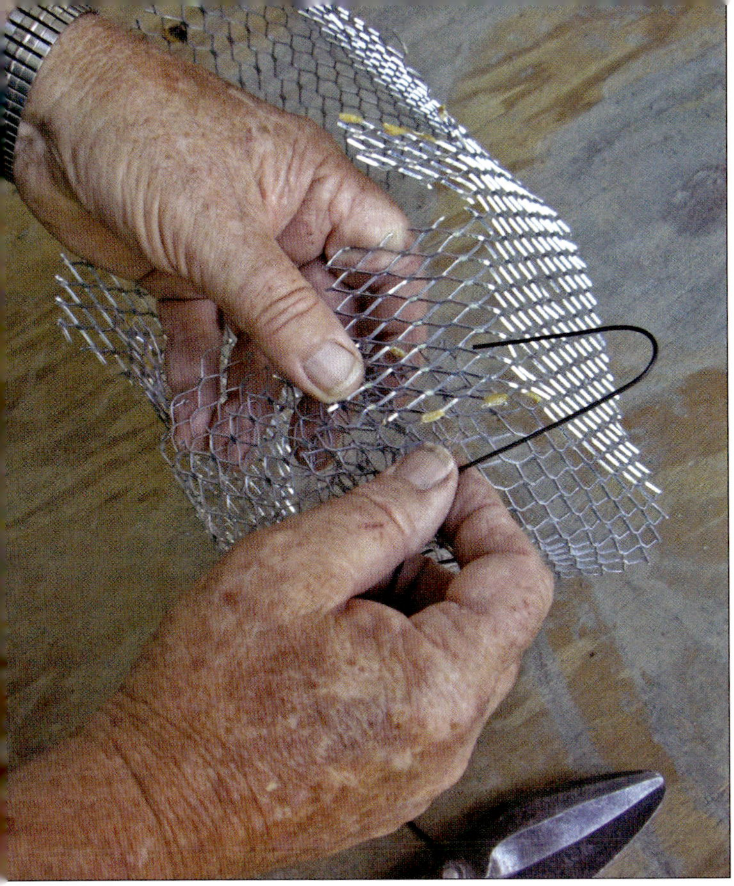

With shears, cut down approximately 2 inches on the top of the lathe, three or four cuts, then fold lathe down. In your mind, you are now shaping the boots. Make two or three ties.

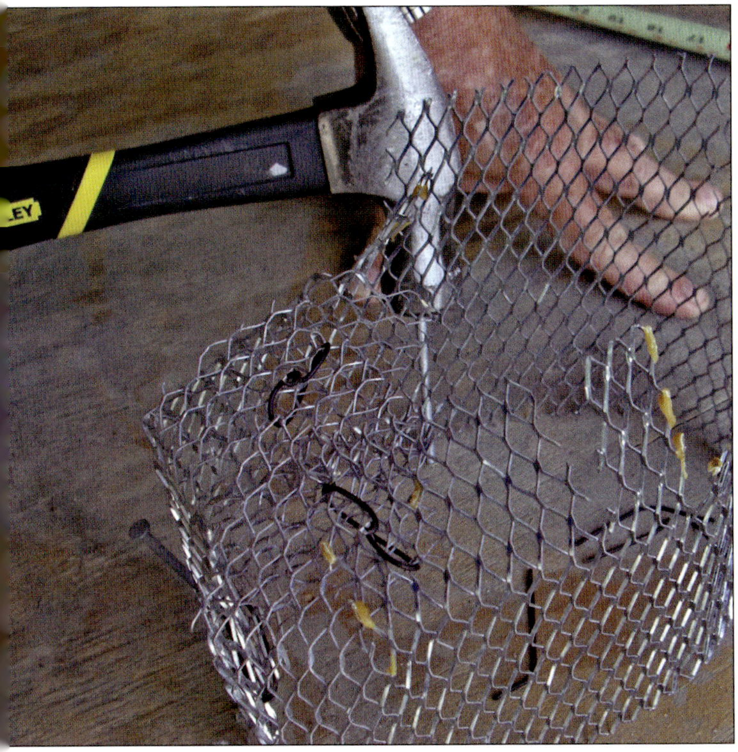

Nail lathe to table.

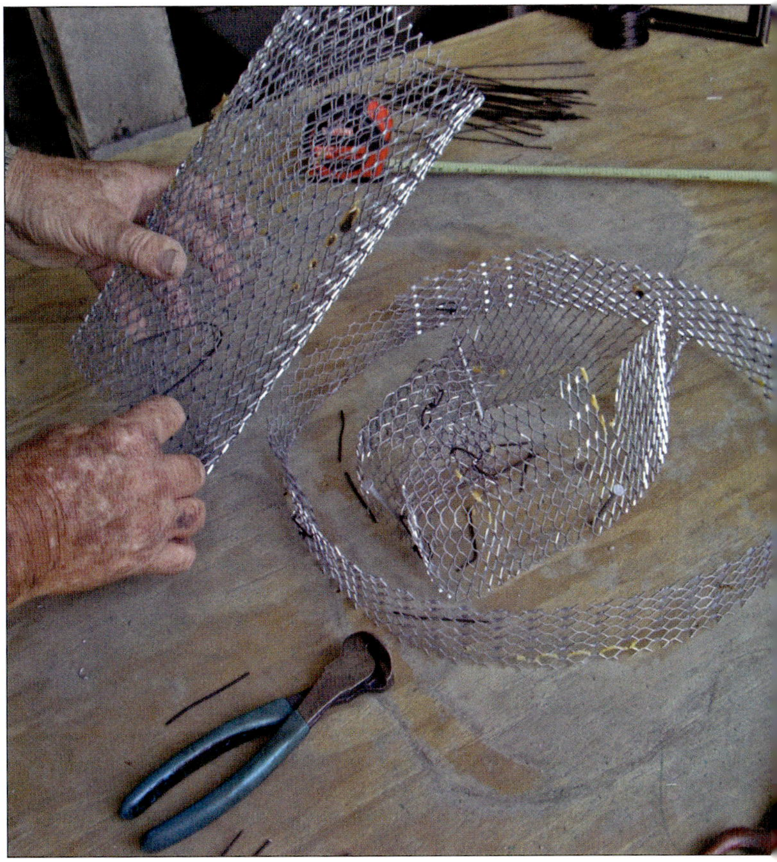

Now take the piece of lathe for the base, put it over the one for the boots. It does not need to be nailed to the table. Cut another piece of lathe, approximately 12 inches x 18 inches. Roll into a circle with a small overlap and tie. When you are creating these sculptures, you have to use your own mind and try to picture how tall you want to make the sculpture, keeping in mind that the lathe and armature needs to be way inside so you can apply and add on mud to shape the piece. This means that measurements that I'm giving you are all approximate.

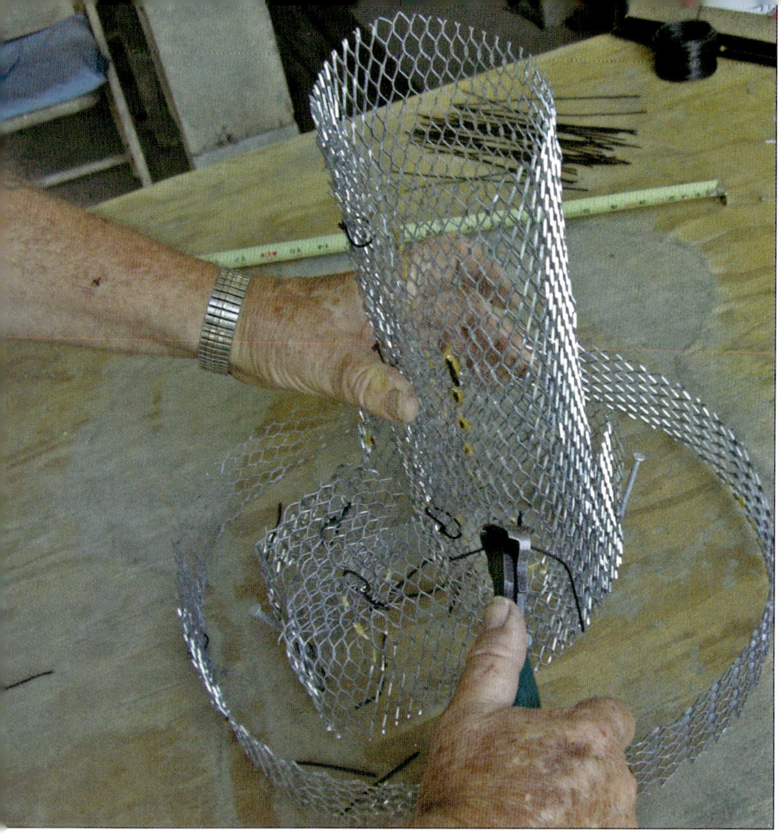

Put this piece of lathe on top of the boots and tie several times.

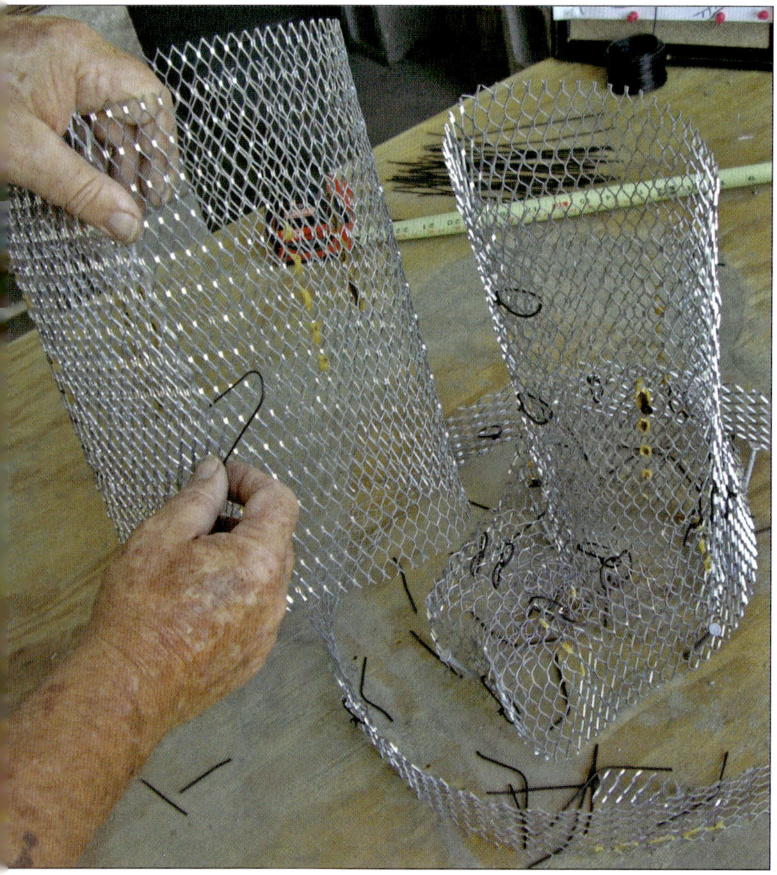

Cut another piece of lathe, approximately 10 inches x 20 inches. Roll 1 to 2 inches of overlap. Make ties.

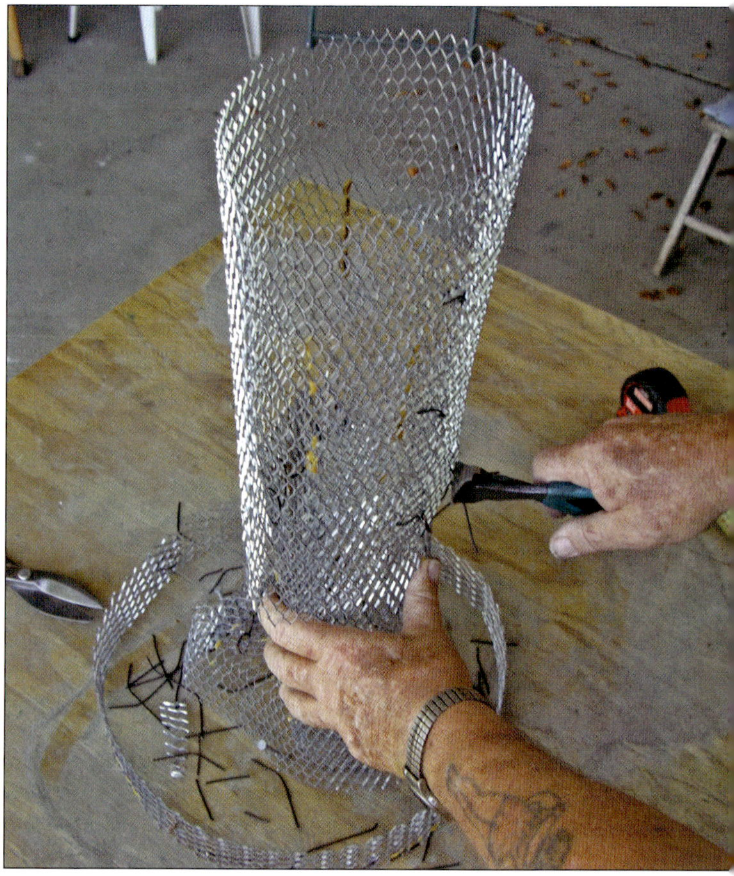

Overlap this piece of lathe and try to judge the height that you want the statue to be by bringing this lathe down or up. Then, tie off.

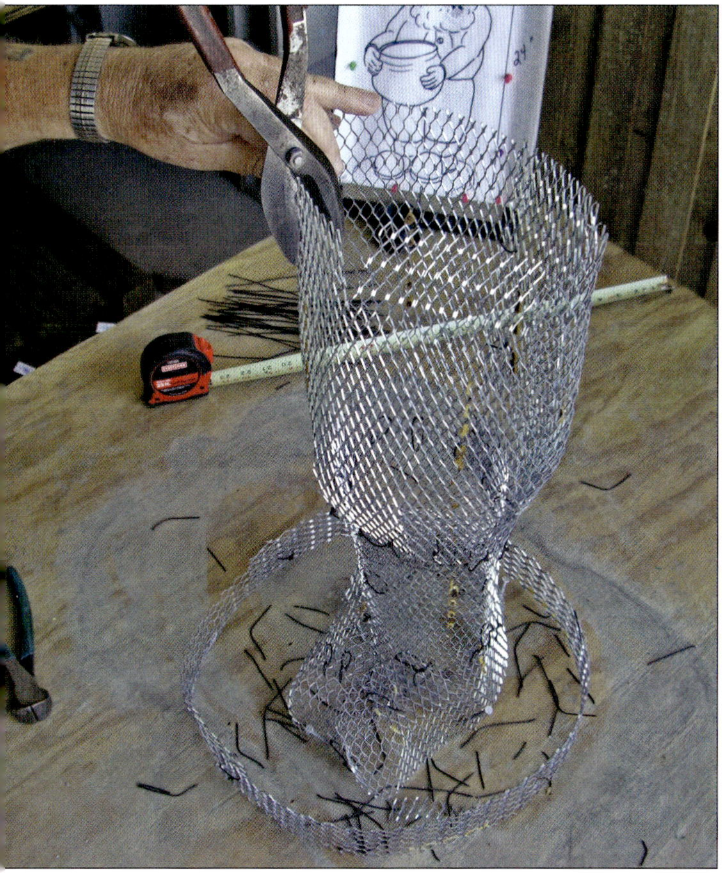

At the top of the lathe, make some cuts down approximately 2 inches all the way around.

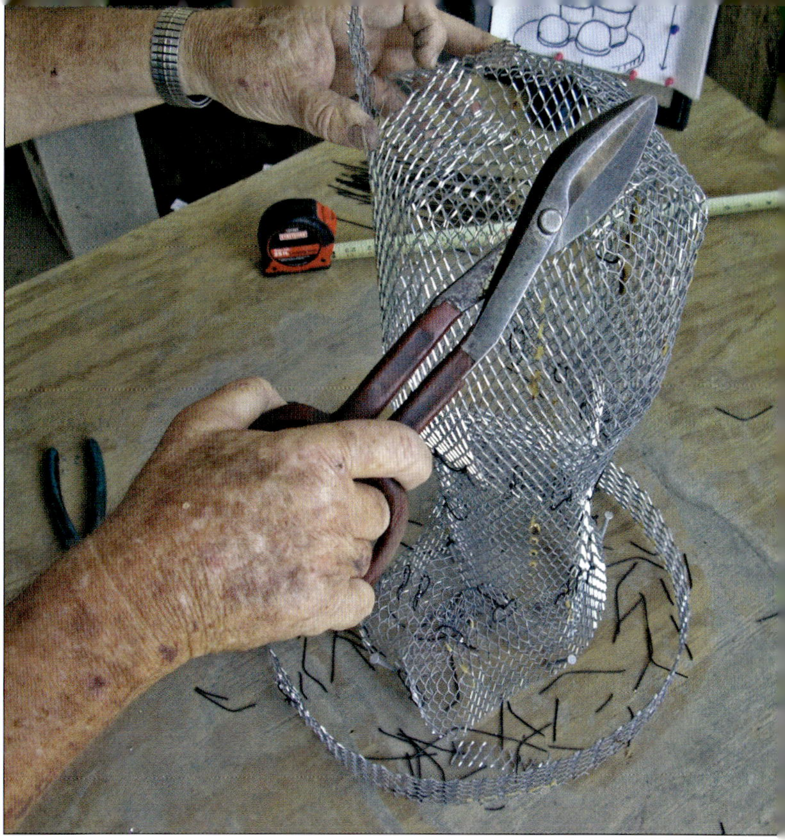

With your fingers, bend and fold in the lathe. Use your shears to tap down on the lathe. This would be the height of the shoulders on the gnome.

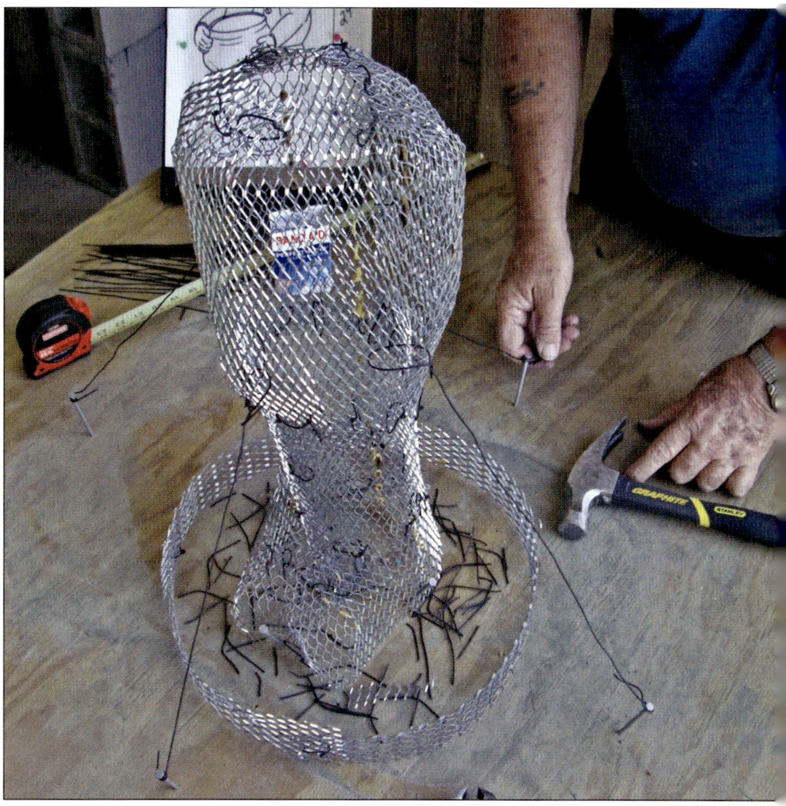

At this stage, you will need a good amount of support for the statue. Nail four nails into the table about 10 inches away from the statue, all the way around. Now cut pieces of tie wire long enough to tie to lathe and twist around each nail.

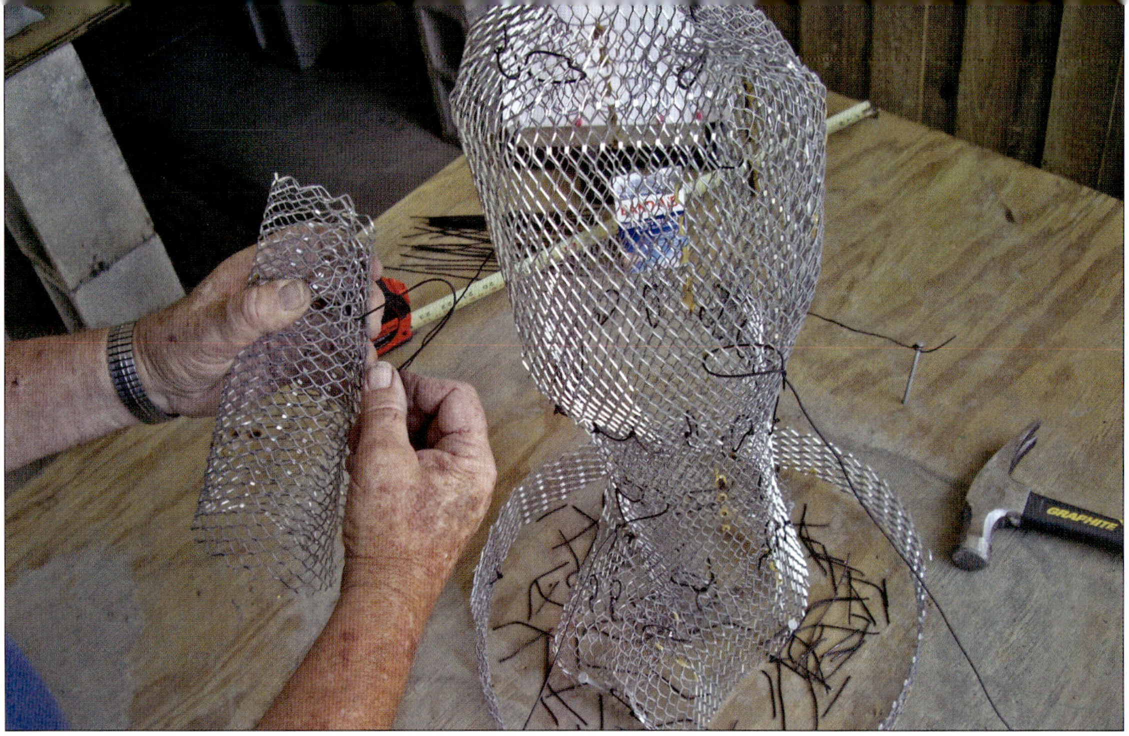

You need to cut a piece of lathe that will be used for his hat and head. Again, the lathe measures approximately 10 inches by 12 inches. Roll and start bending and shaping the lathe. Tie lathe together, then on the bottom, make one-inch cuts up all the way around every two inches. Use your hands to just bend lathe out.

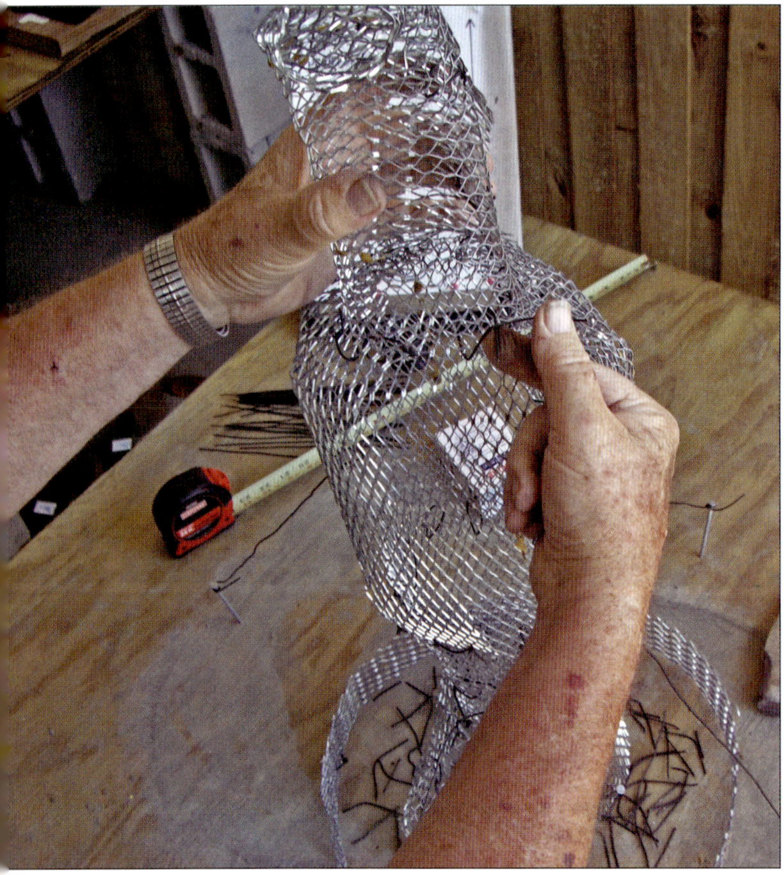

Put the head and hat on the shoulders. Make ties where you've bent the lathe out.

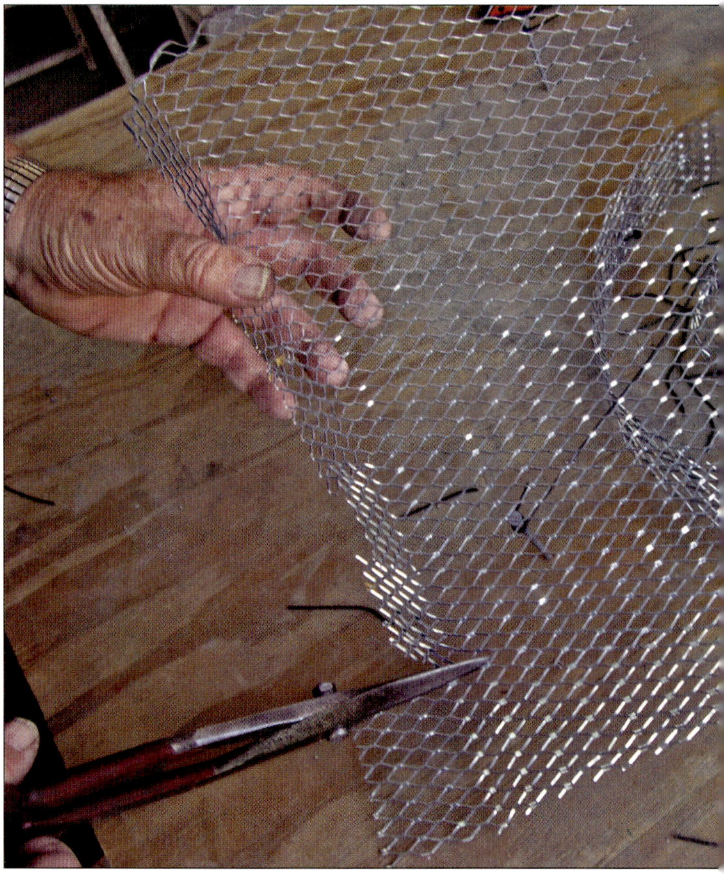

We will now assemble the flowerpot. Cut a piece of lathe, approximately 10 inches x 20 inches. Before rolling, make 2 inch cuts, 2 inches apart, around the bottom of this piece of lathe. Now bend them inward.

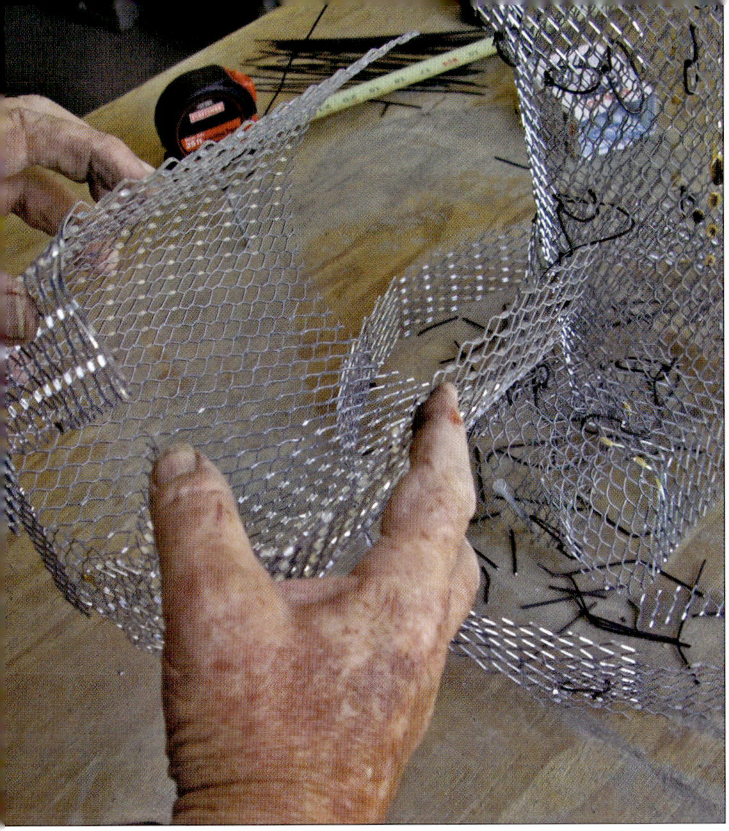

Now, roll the wire, and you will see the bottom close up.

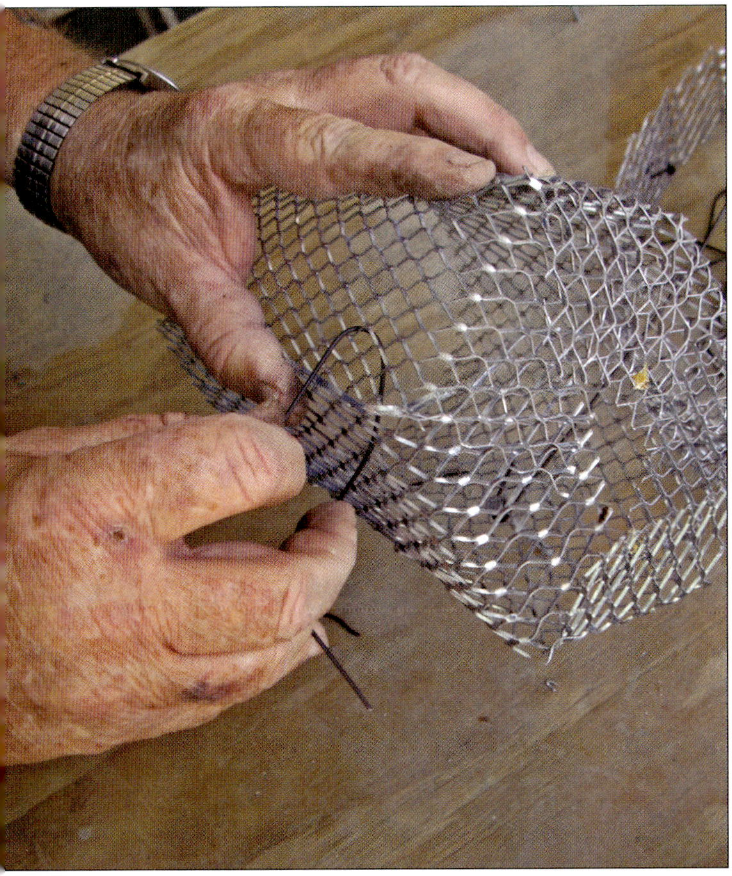

Overlap. Send wire through lathe.

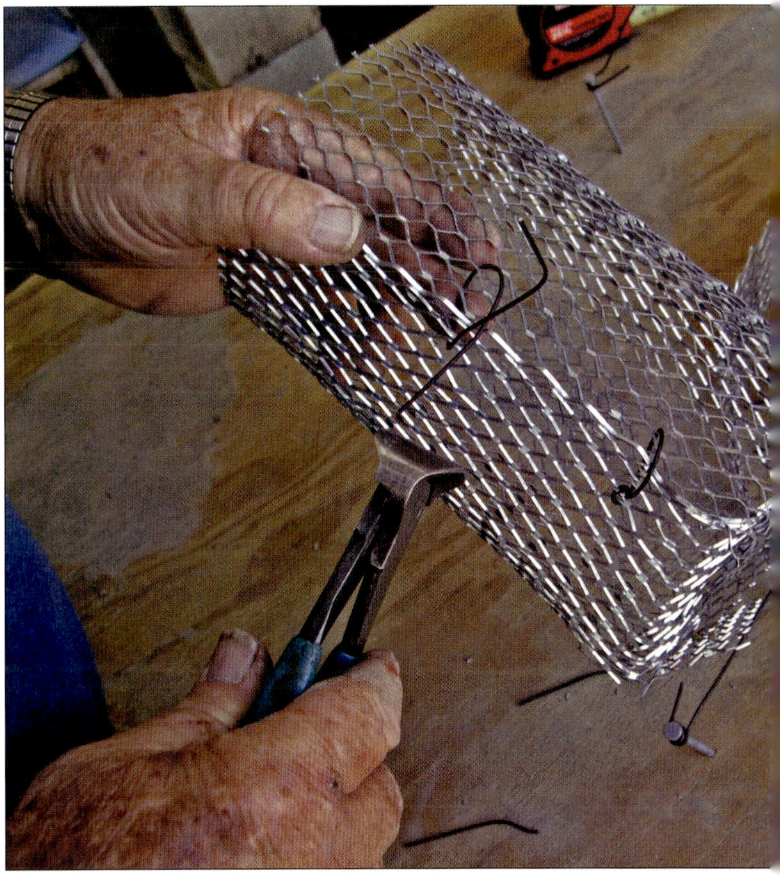

Make a twist and cut off.

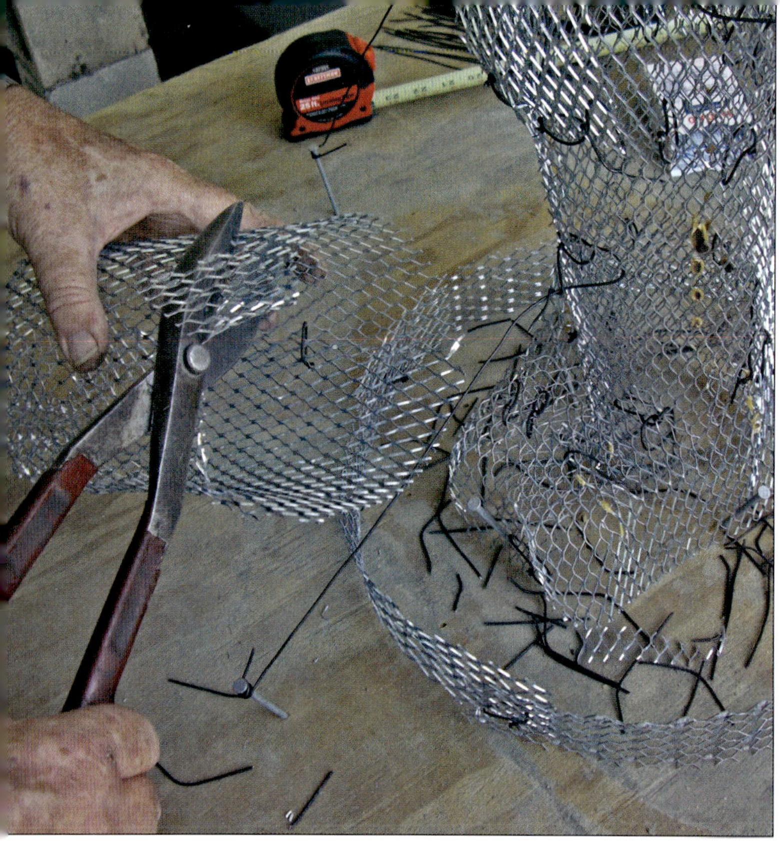

Tie bottom together. Trim off any excess, making the top of the pot level.

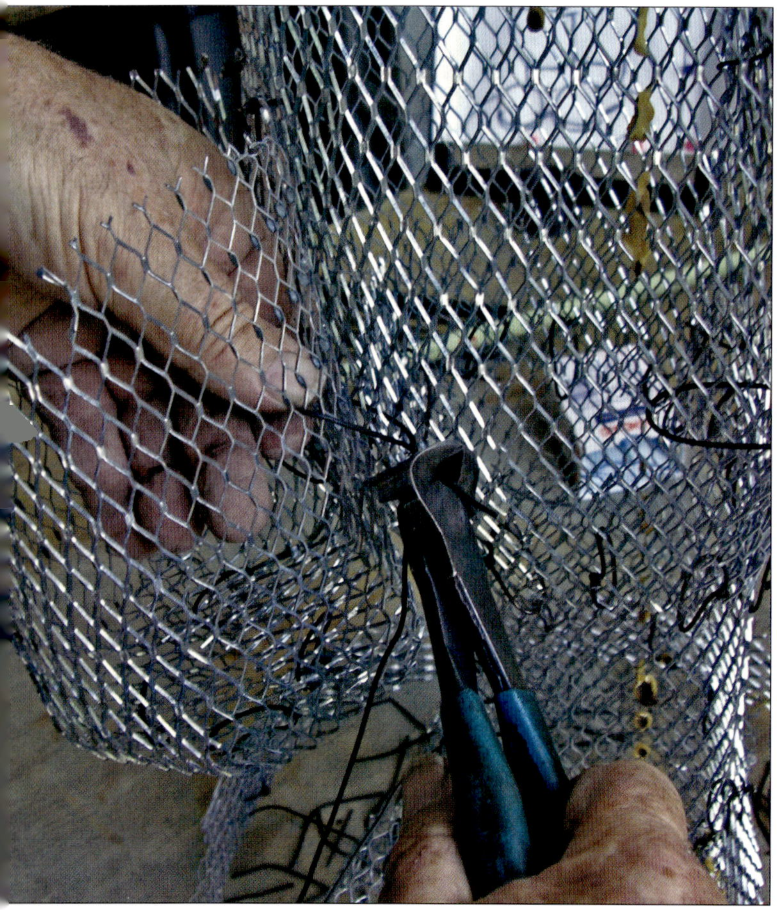

Tie the pot where you think it will look right on the lower part of stomach; remembering, you have to put lathe on for his arms.

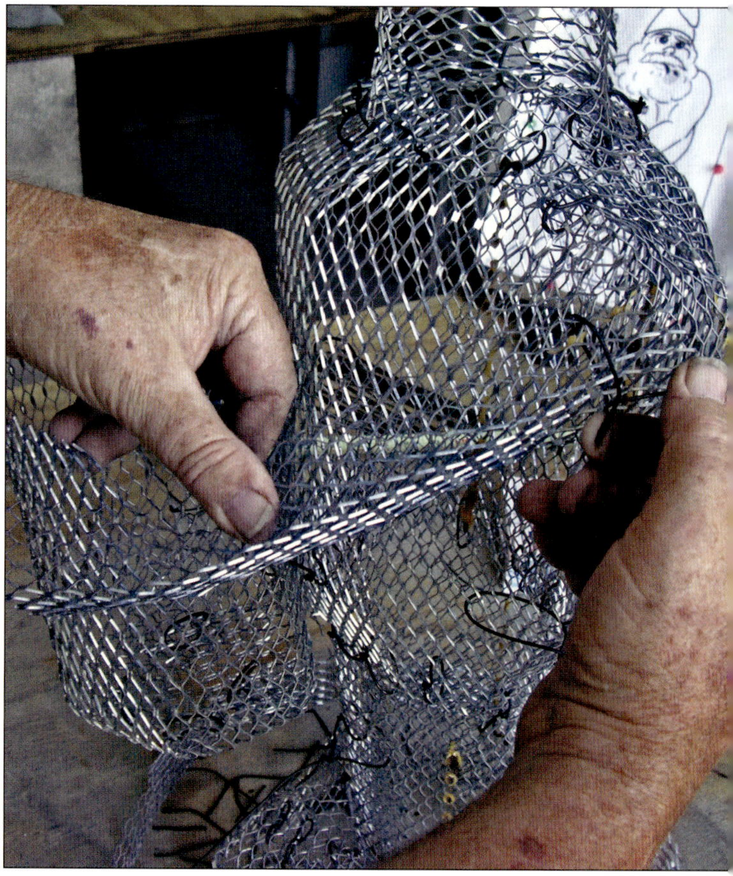

Cut two pieces of lathe approximately 3 inches x 10 inches. Bend into a U shape. These will be his arms and the open end of the U is so you can make the arms solid. Squeeze both ends of the U shape together.

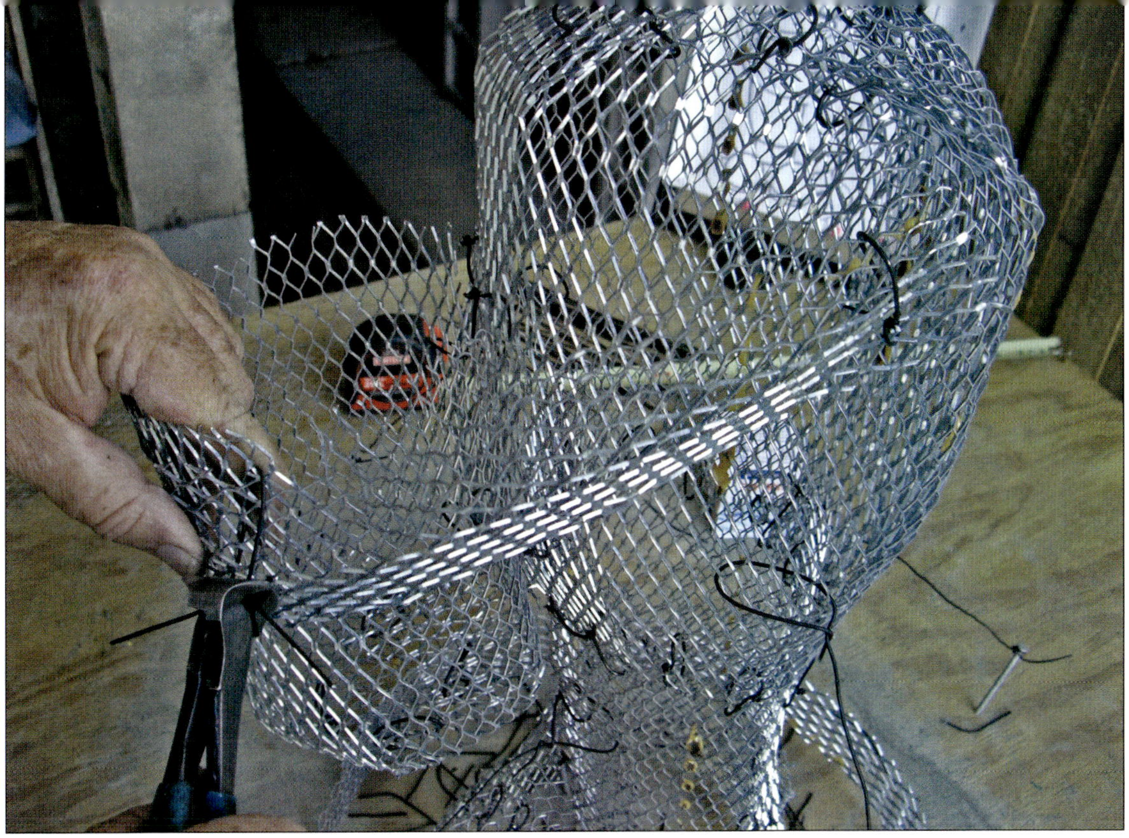

Now, tie one end of this lathe to the shoulder, and one end to the flowerpot.

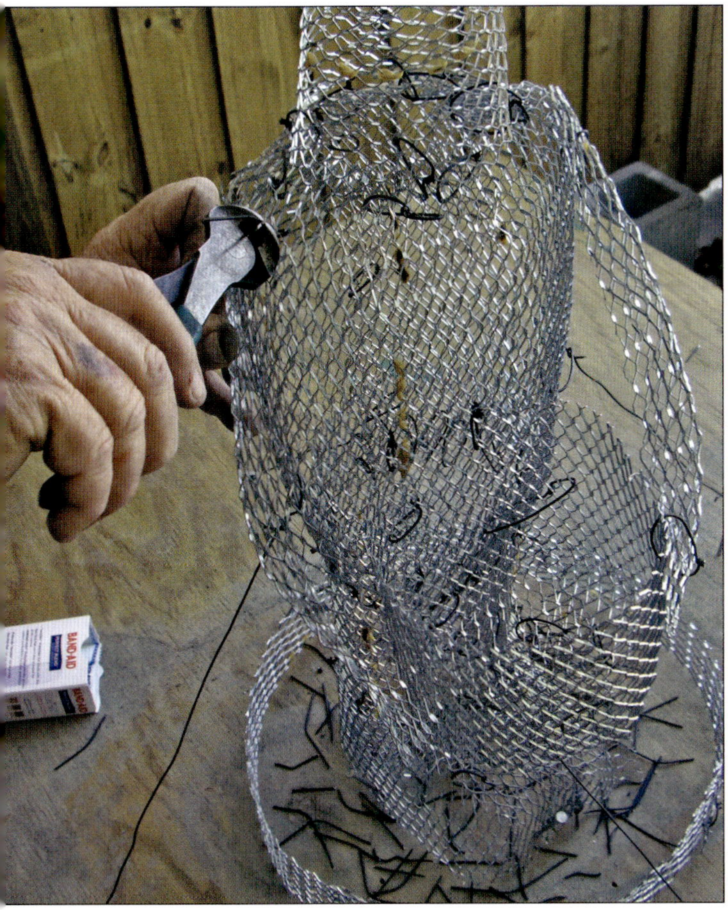

Repeat this process for the left arm.

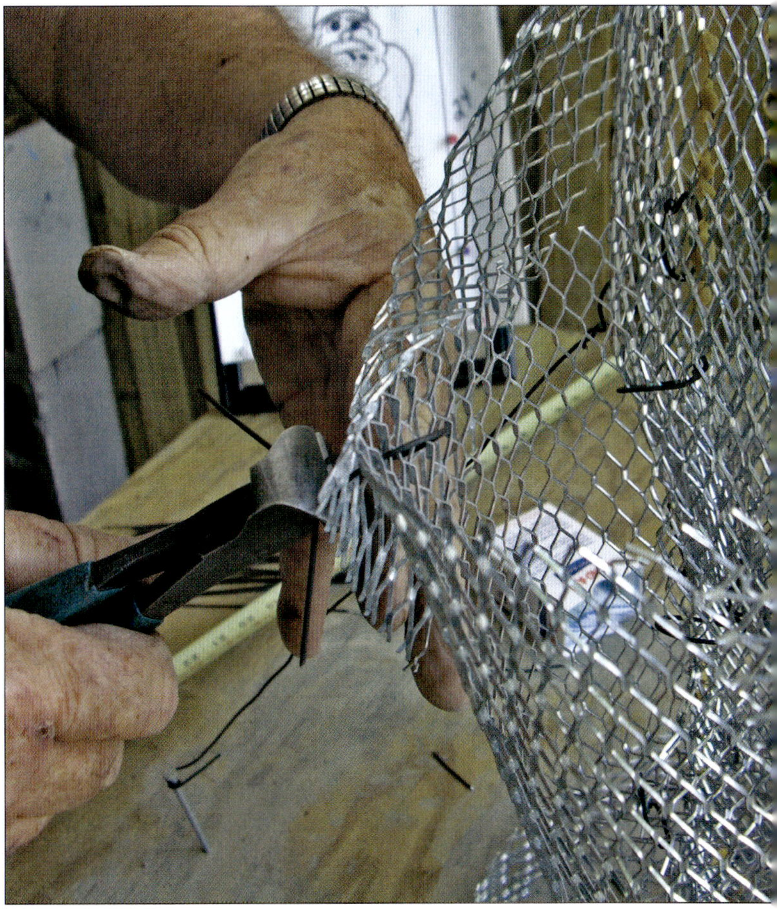

Same.

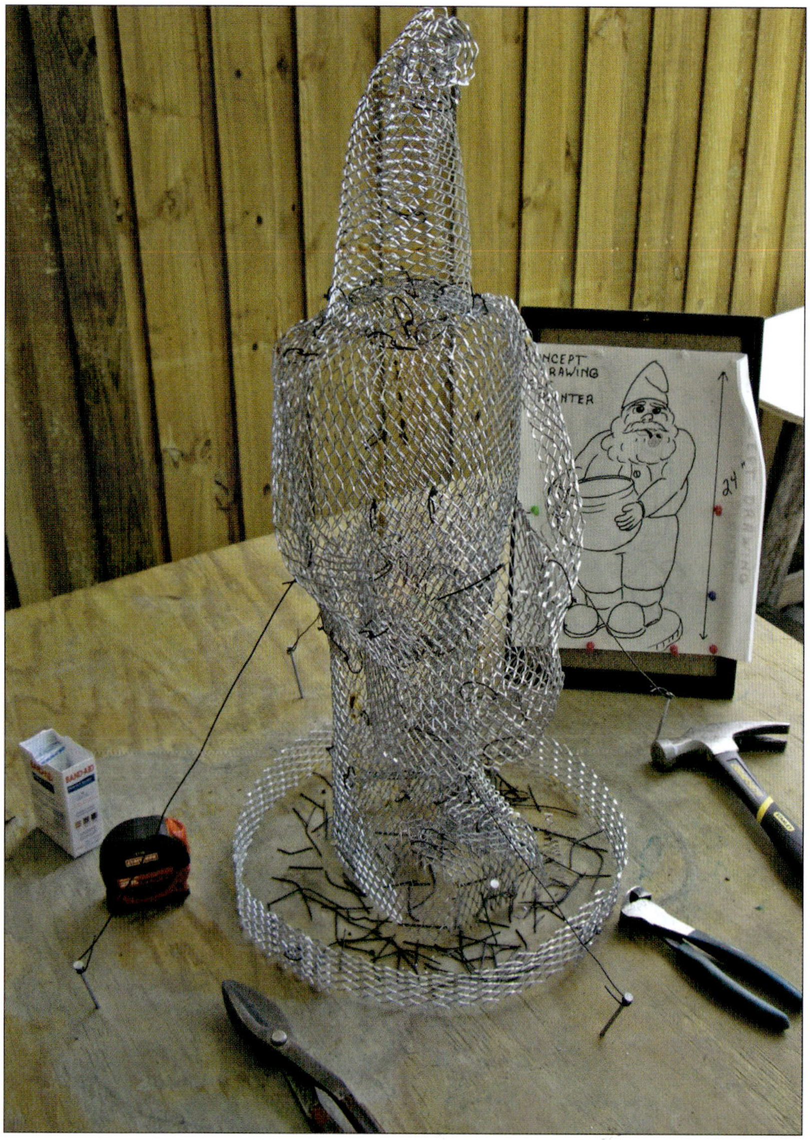

Ready for mud.

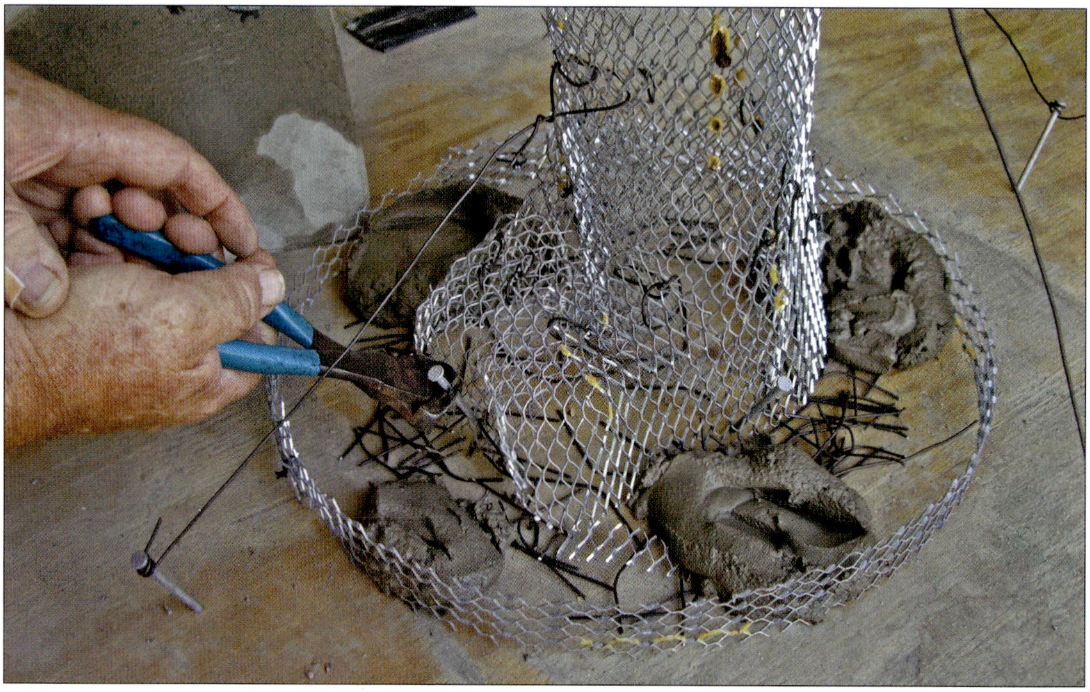

Put some mud around the boots, then pull out the nails.

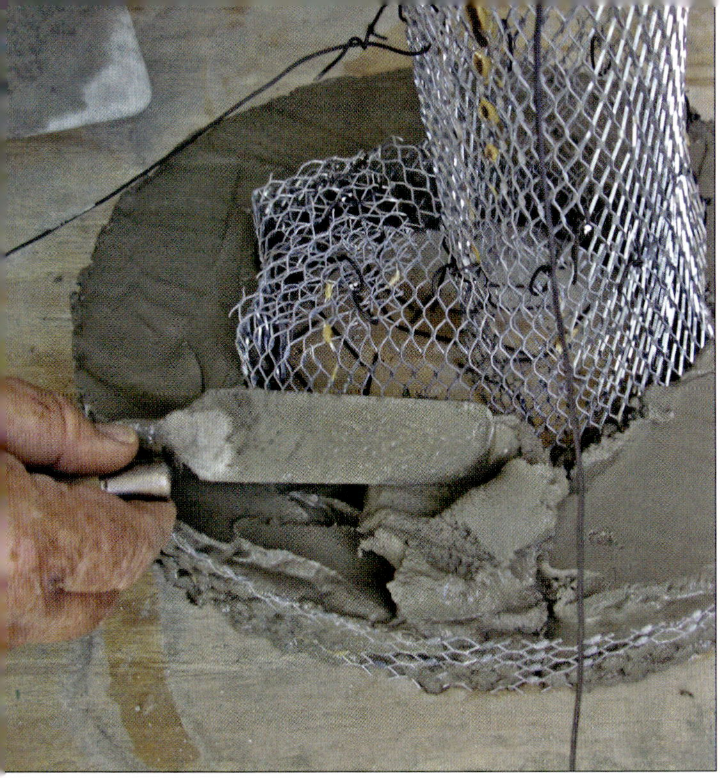

Fill in the base wire solid with mud.

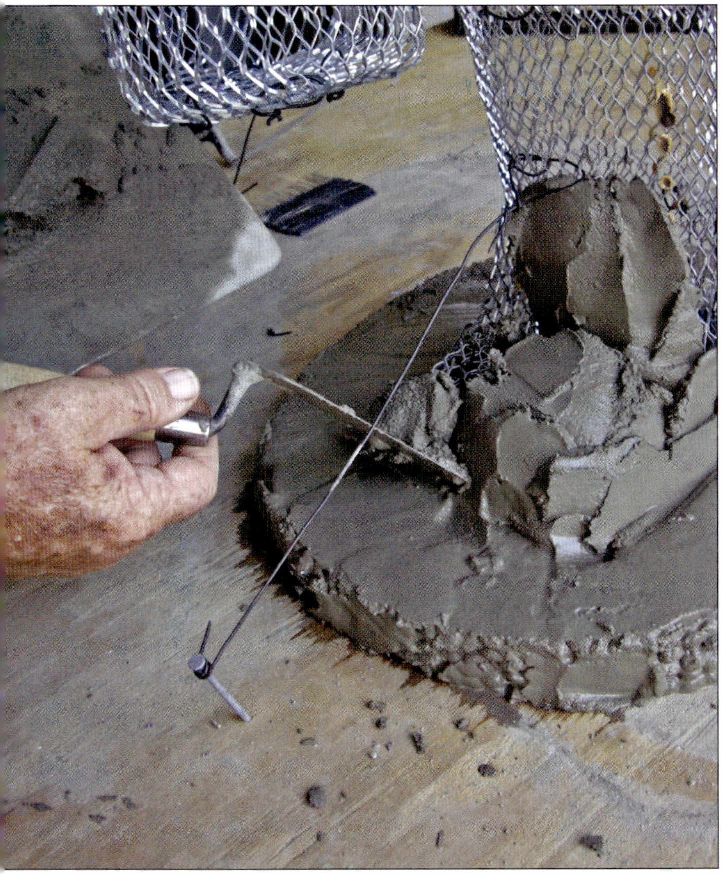

Push mud through the lathe to make boots and a portion of the legs solid.

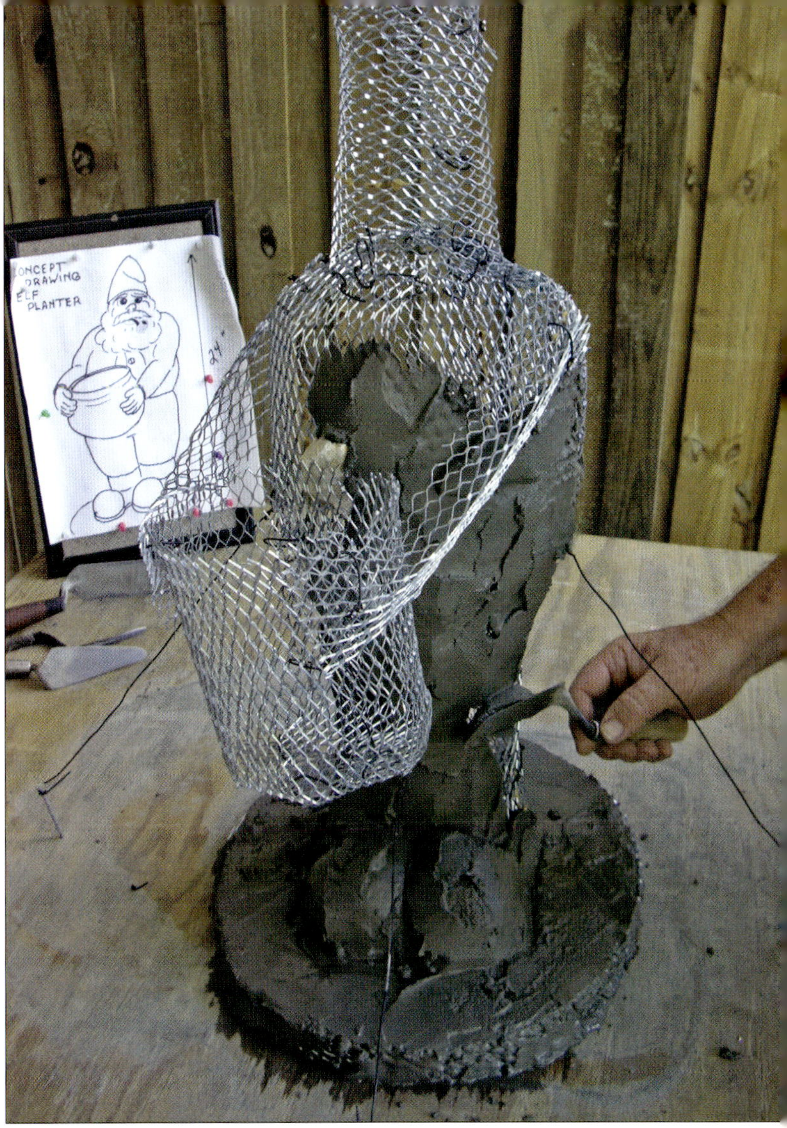

Start mudding from the bottom up, not from the top down.

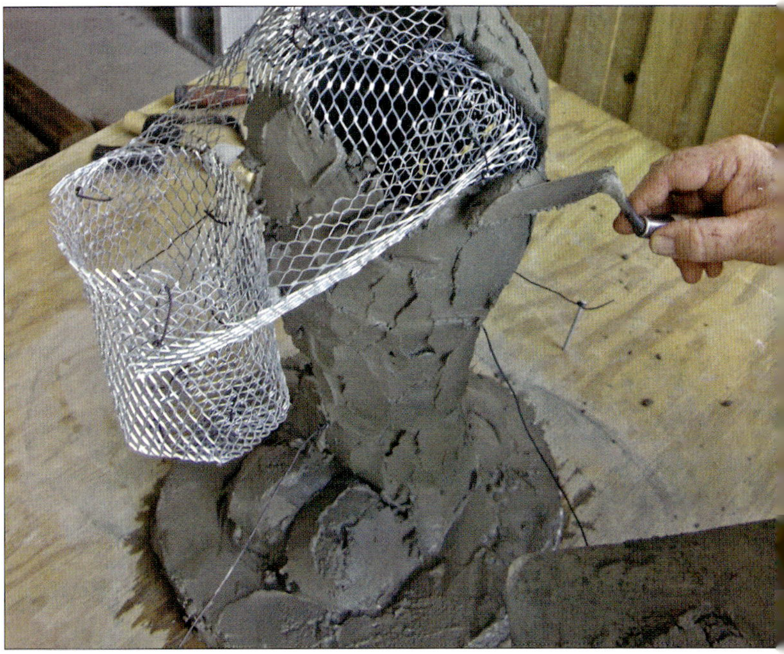

At this stage, all you're trying to do is get mud all over the armature.

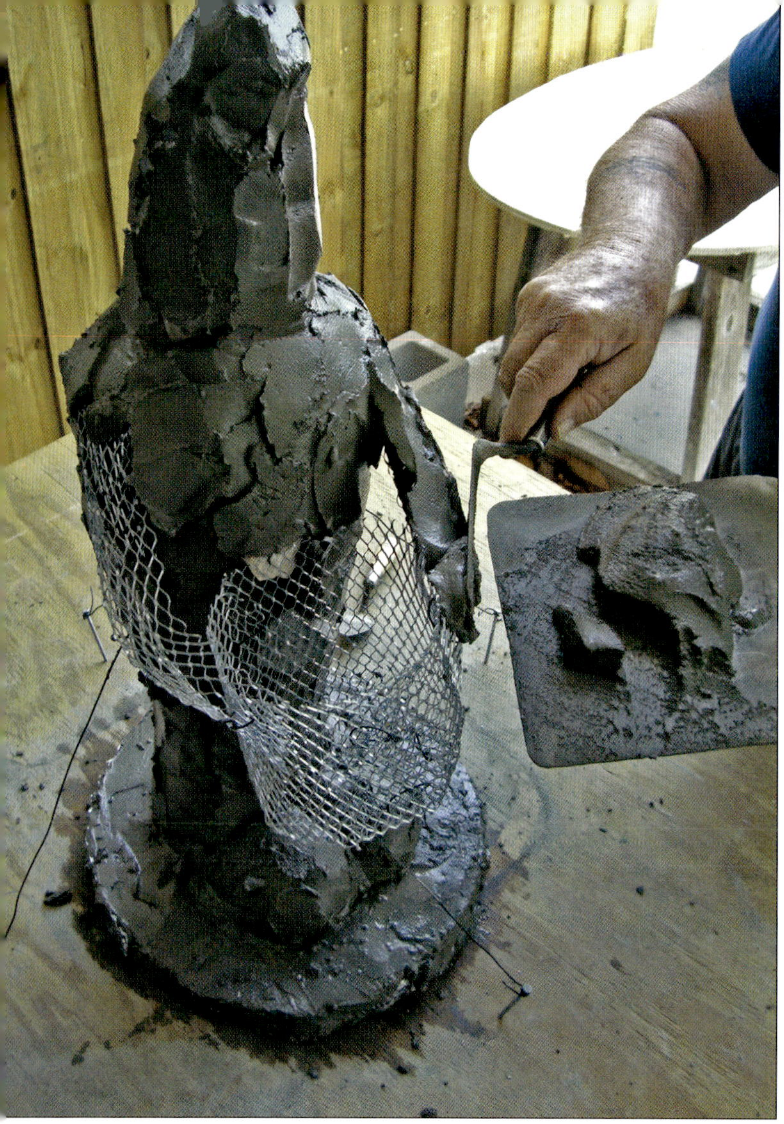

It does not matter what the figure looks like at this stage.

When this mud sets, in two days, it will become easier to shape because then you can press hard with your trowel.

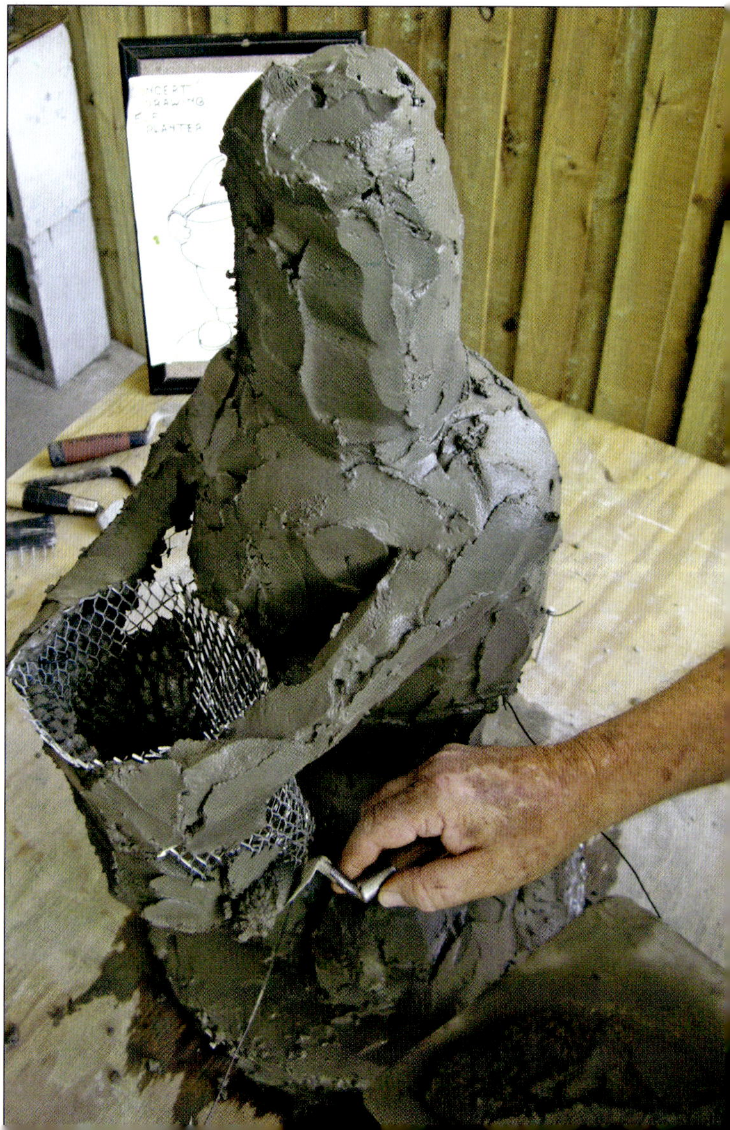

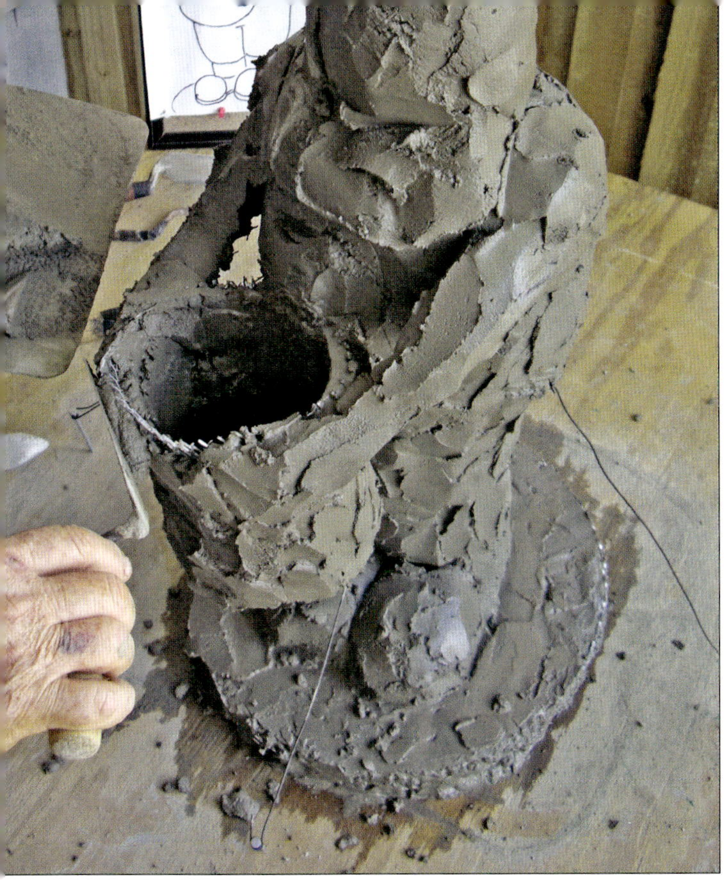

Finishing the first coat.

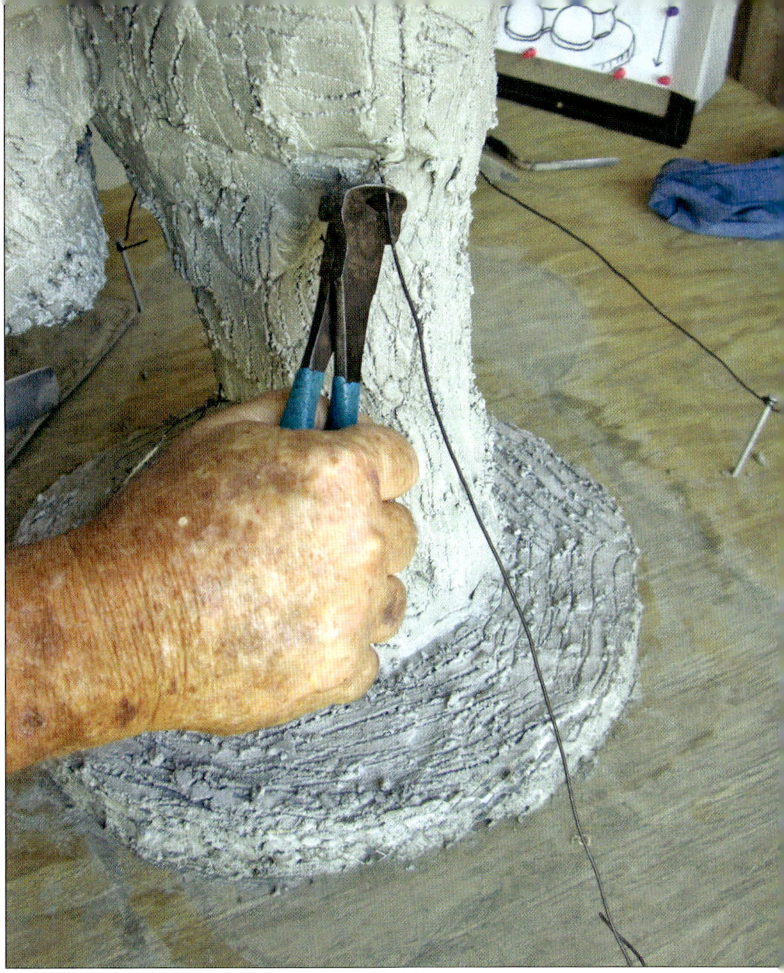

When you first mix concrete, it's soft. In one day it's hard. In two days, it's five times as hard; so, in two days you are ready to cut off the support wires.

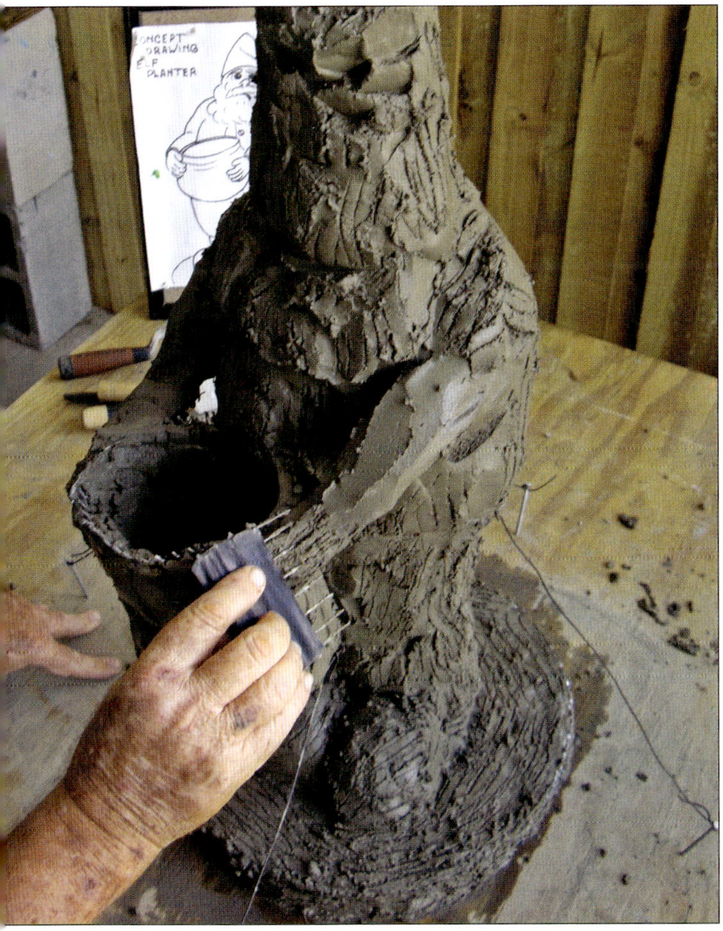

Using scratch wire, try to scratch and make rough everywhere.

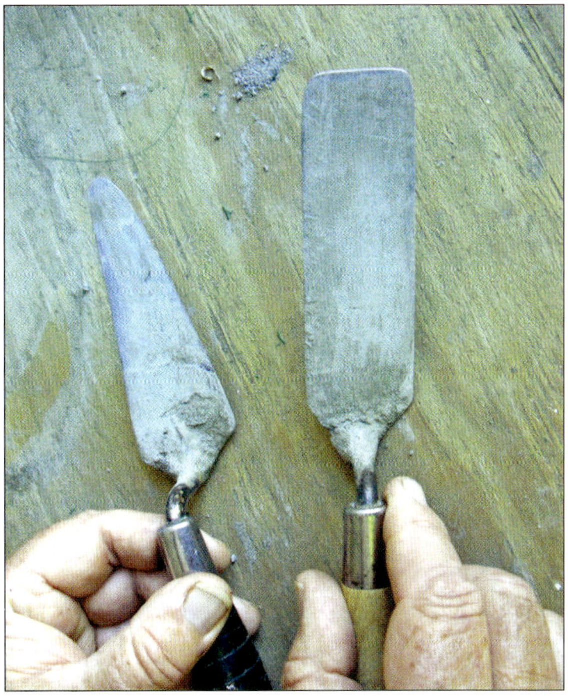

I wanted to show you these two margin trowels. The one on the right, new, the one on the left, worn down from maybe two years of use.

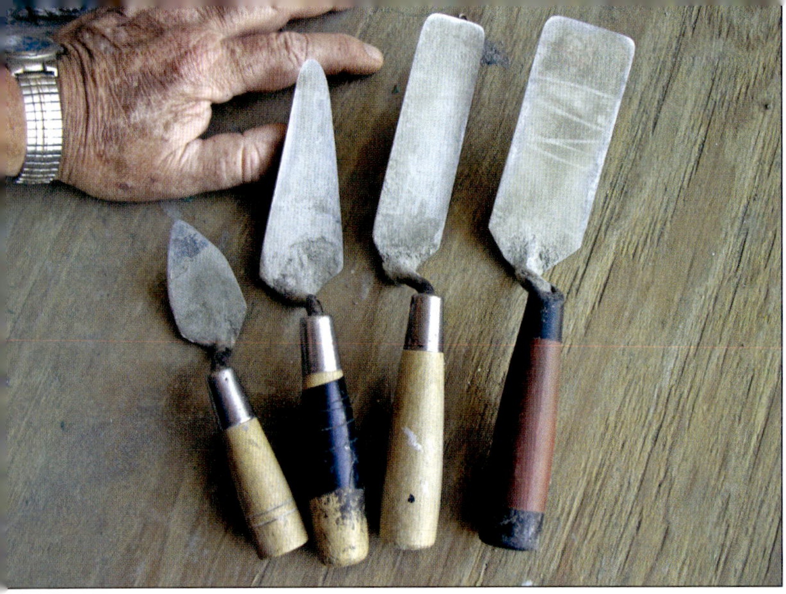

Same thing with all these trowels. There is a way to make your trowels like these; it's called "grind them down." I've never done that, so I suggest you buy a point trowel and two different sized margin trowels and start with them.

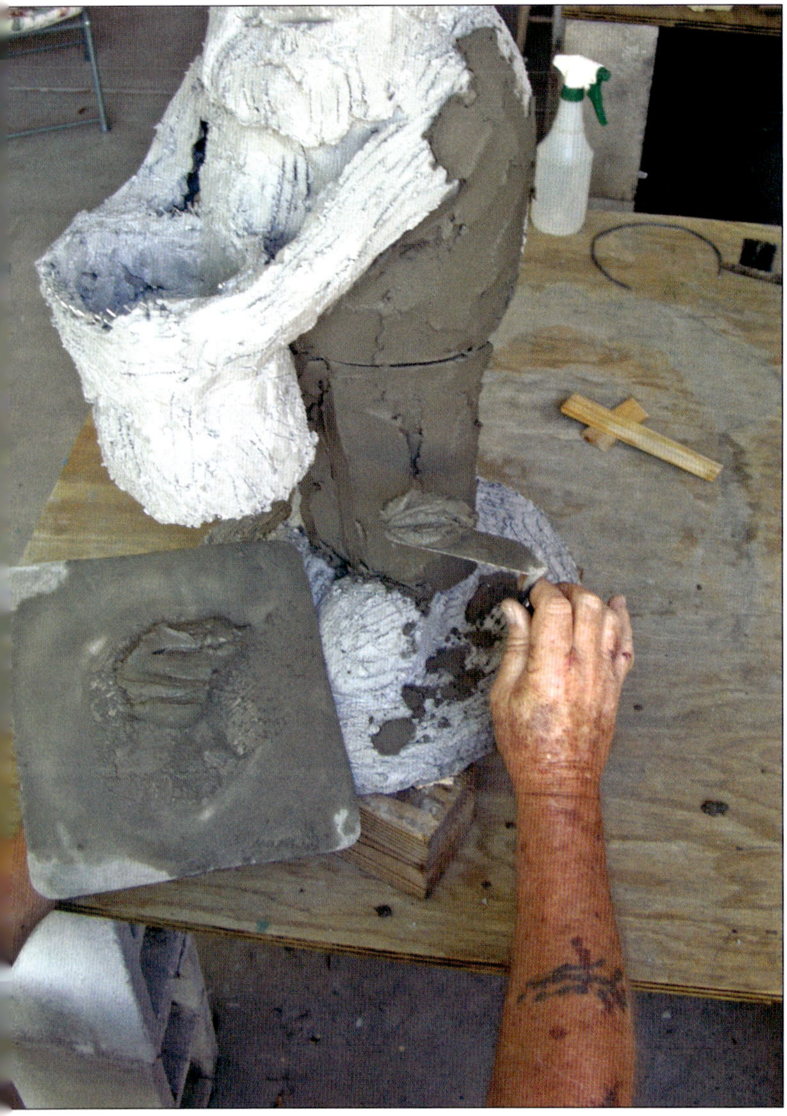

You can start mudding now from the ankles up, because the boots and base are solid.

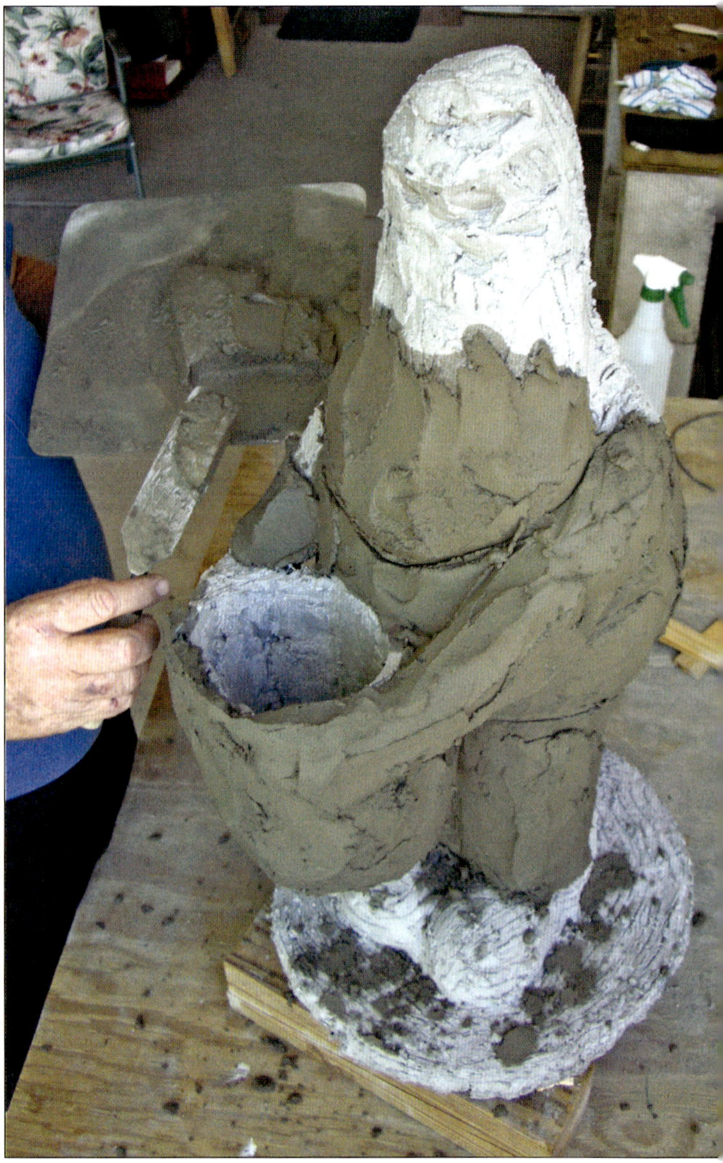

Now, you can go to different areas and start building them out; for instance, the beard, the arms, and pot, which need a lot of mud for the shape.

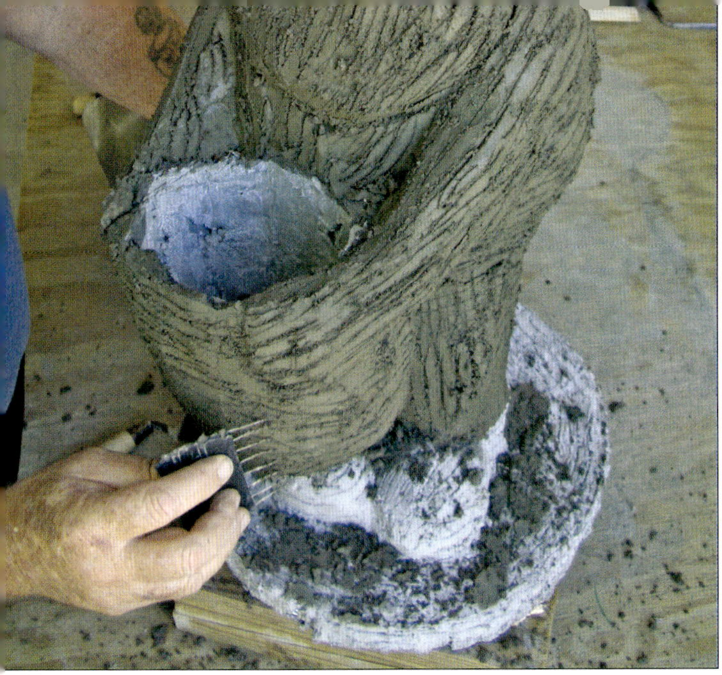

Again, if this mud starts to dry out, scratch with your scratch wire to make it rough and do not worry about cracks, because no matter how many come along, you will be covering them.

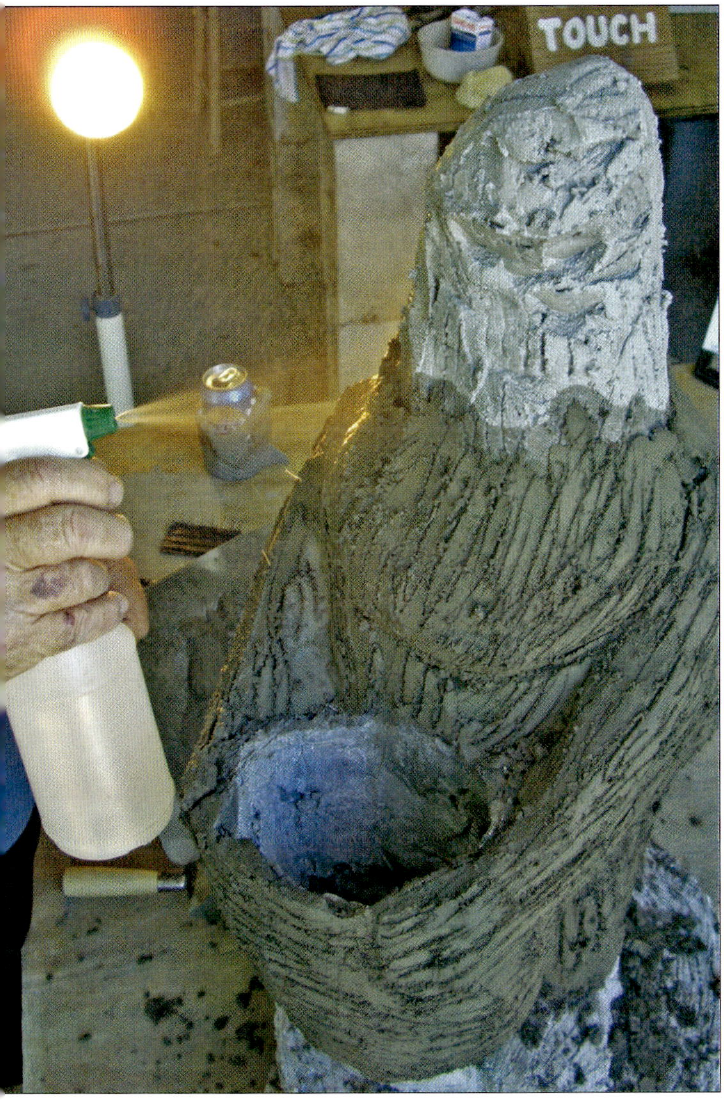

You also must have a spray bottle on hand all the time.

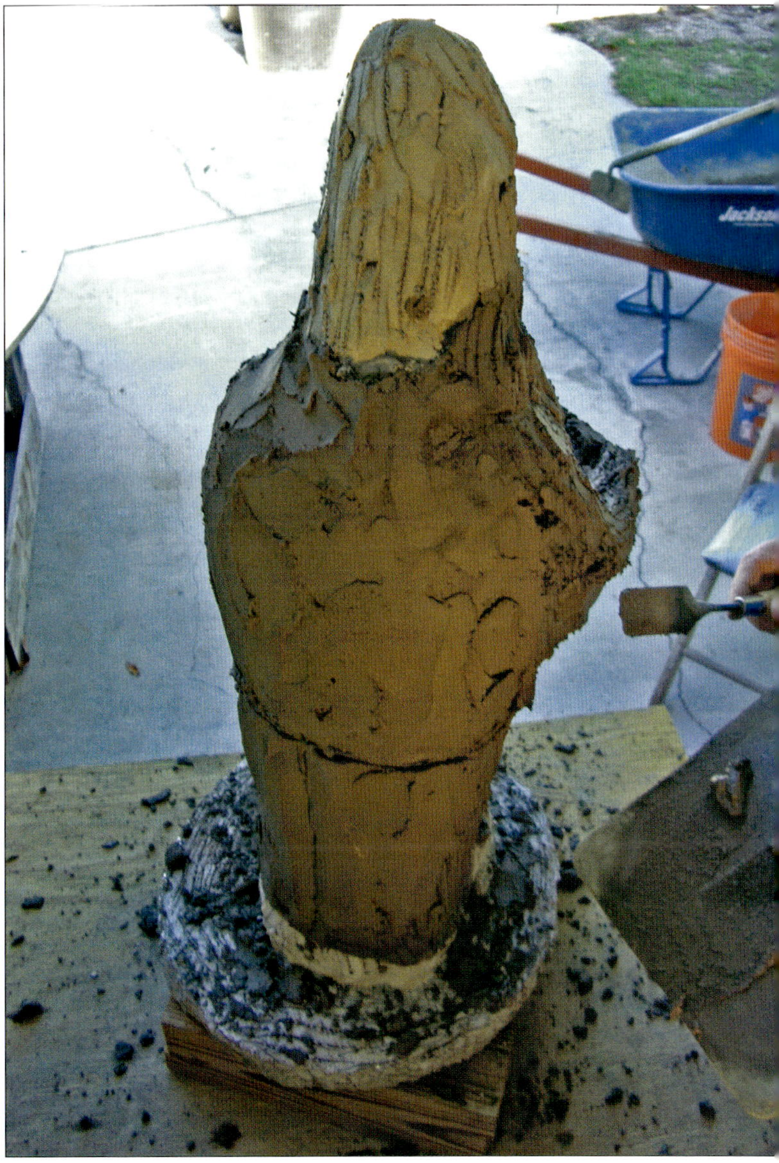

Shaping the back area.

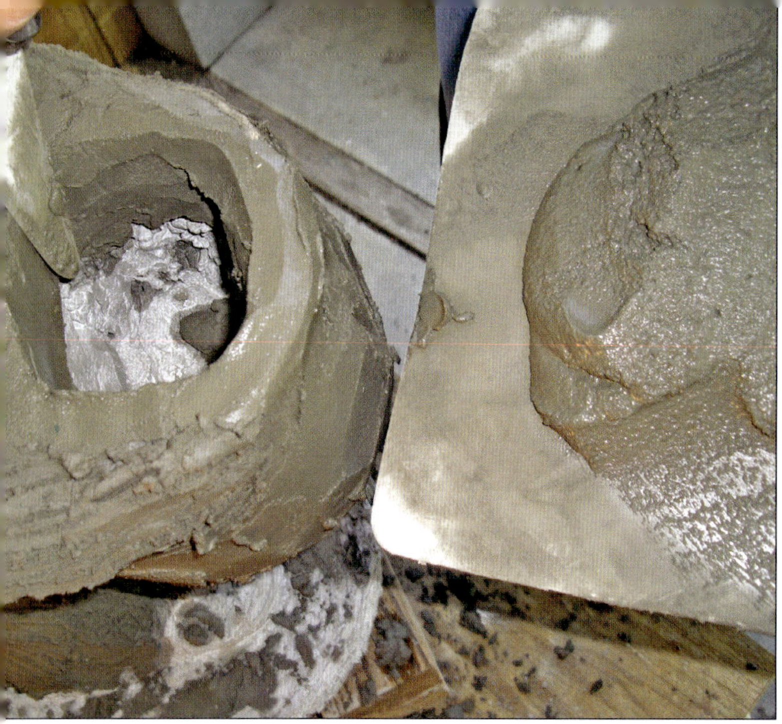

Mudding inside the pot. You do not have to go all the way down with this coat. Also, on the first coat, I made a hole on the bottom of the pot for drainage. I left a piece of wood stuck in the hole so it would not refill with mud.

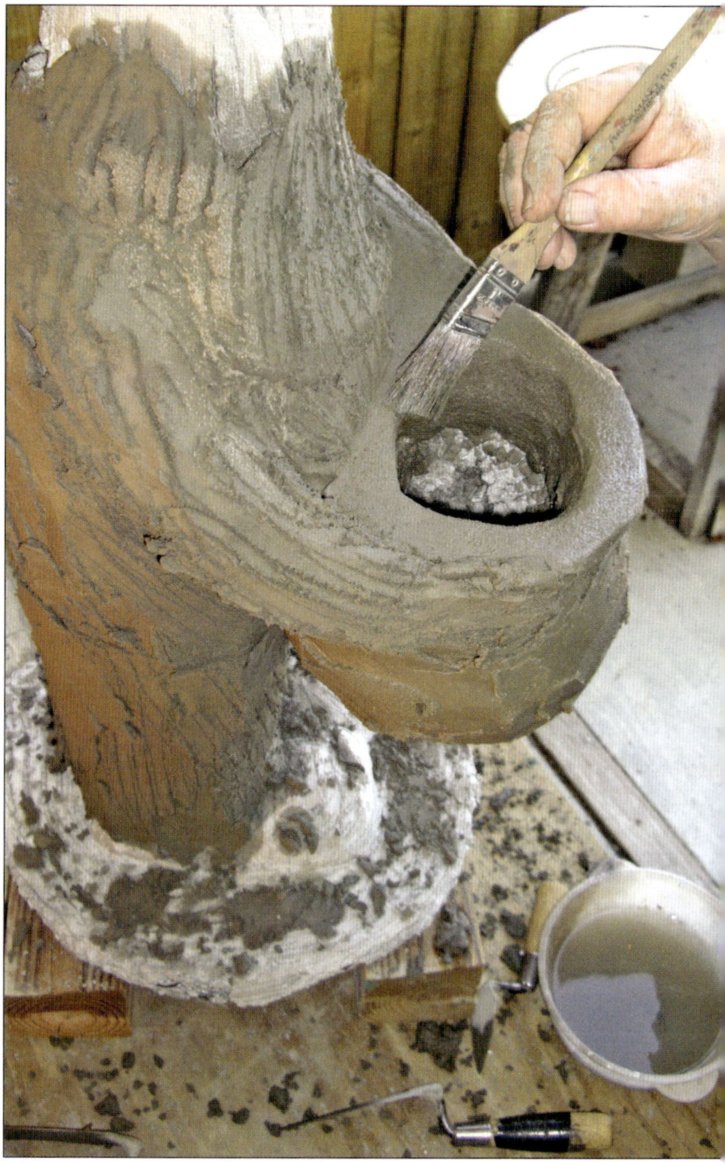

If you cannot get your hand and the sponge inside the pot, you can always just use a 1 inch paint brush to do the same thing.

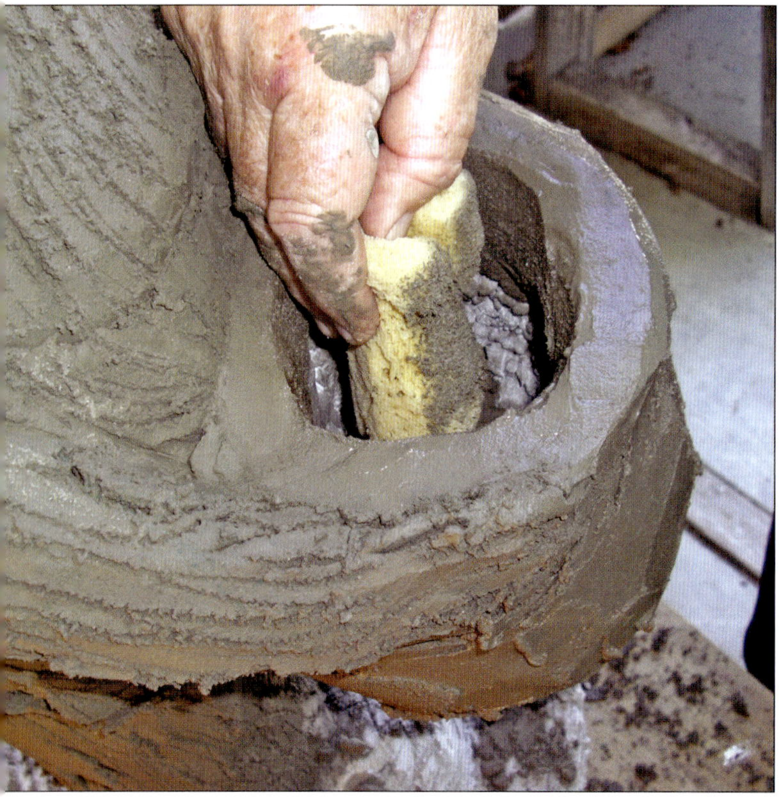

The inside of the pot can be left rough. Take a small piece of wet sponge; float the inside of the pot. This, of course, will give you a sand finish. To accomplish this, just take a damp sponge and, when the concrete is set and you can put your finger against it and it does not leave an impression, it is time to sponge float. Using circular motions with a damp sponge, what you are doing is bringing the sand up to the top and creating a sand finish.

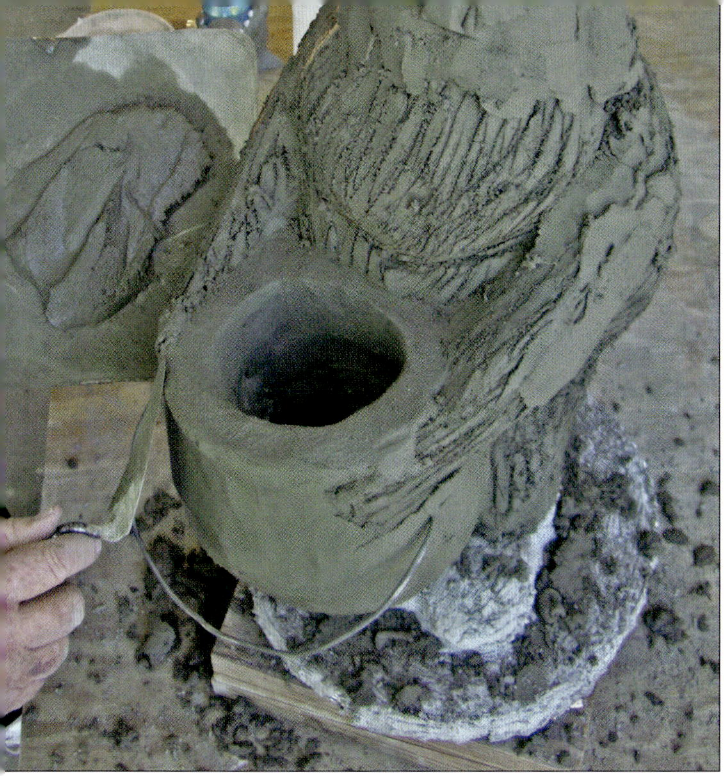

More shaping.

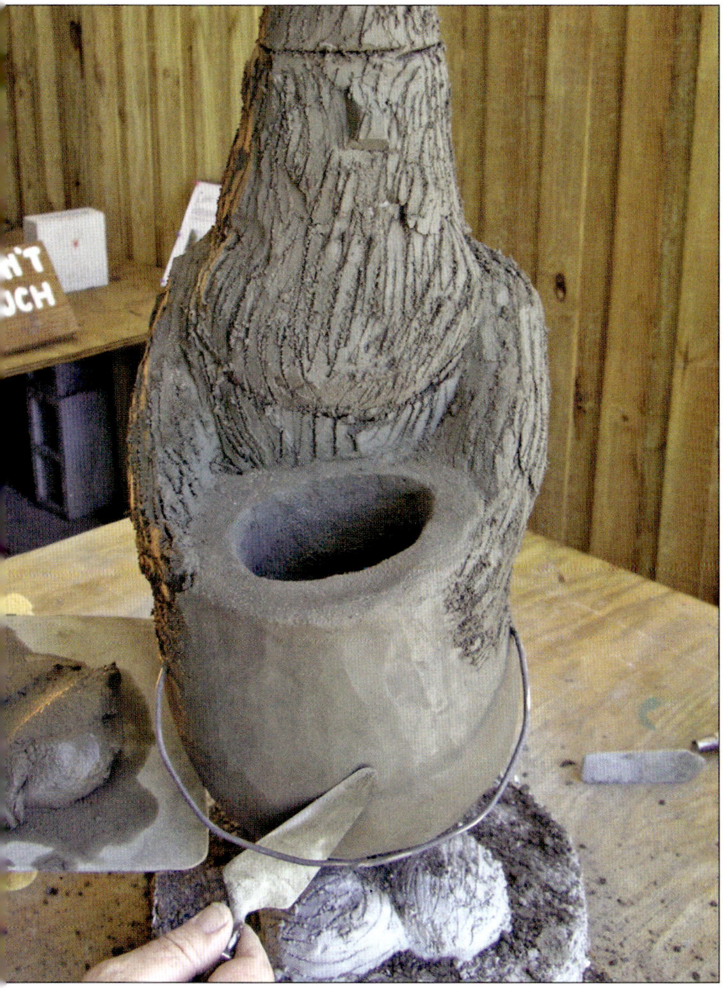

I took a piece of 1/8 inch thick galvanized wire, bent and shaped it for a handle on the pot. Both ends of the wire were bent in approximately 1/2 inch to set in the mud.

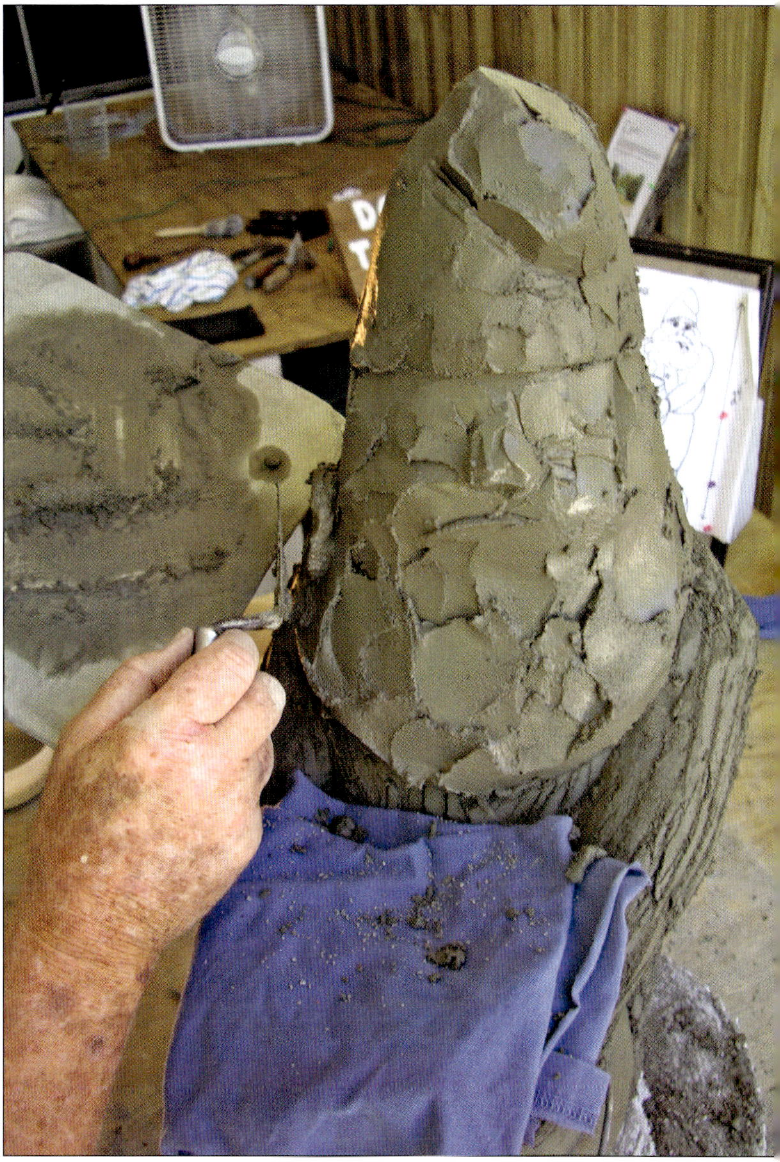

I put a rag over the pot to stop mud from falling inside it.

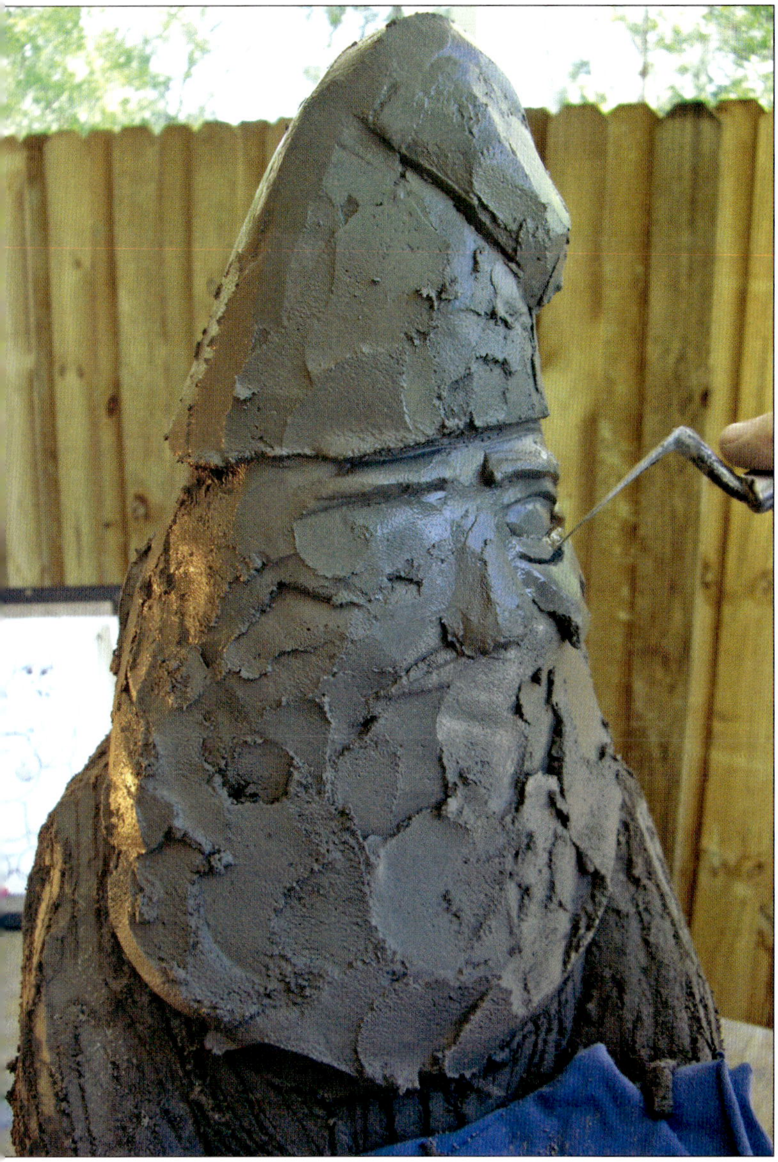

We are now working on the detail.

Same.

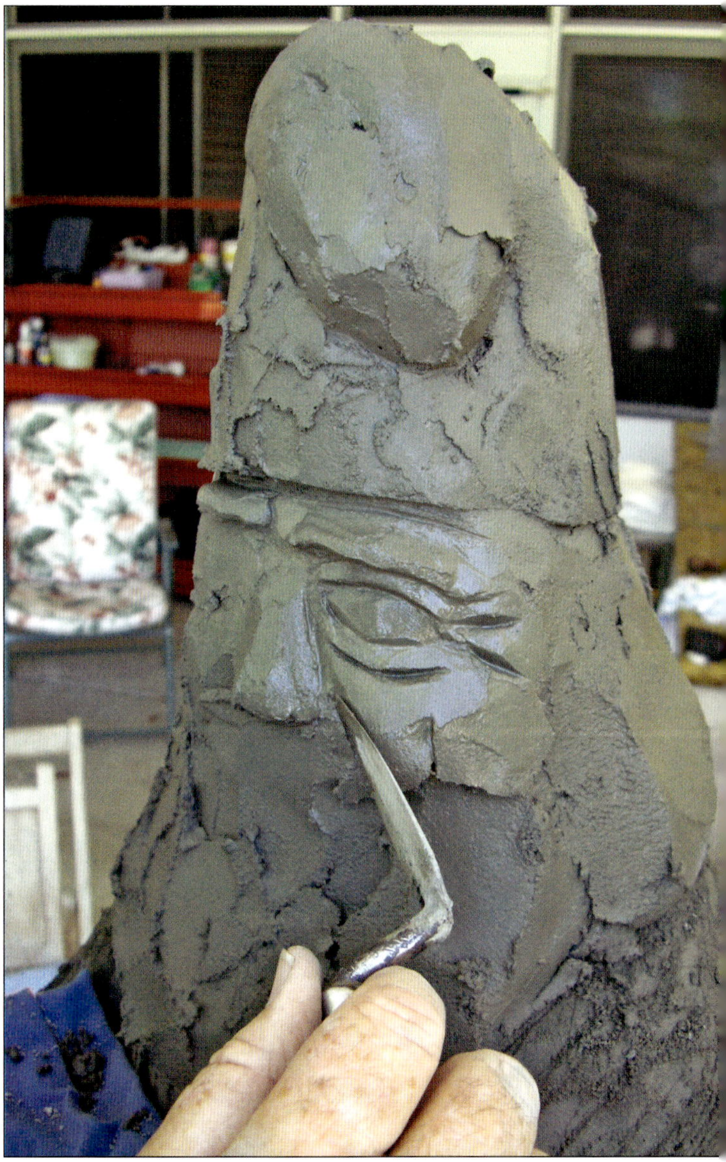

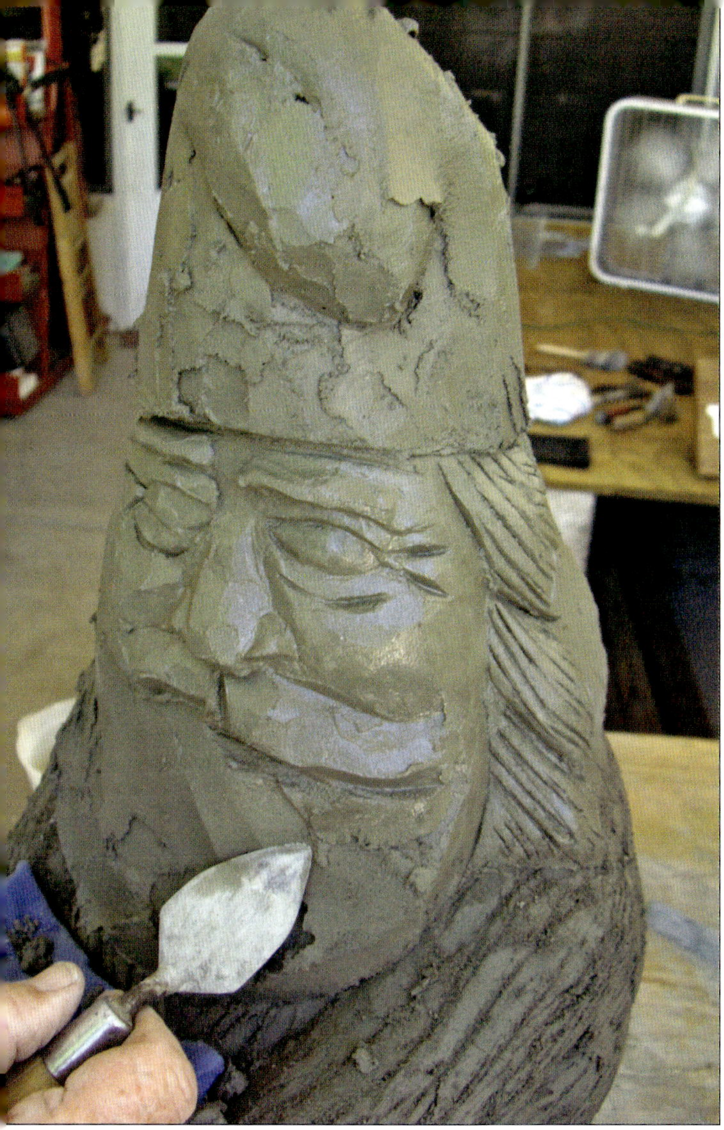

Do not be afraid to stop at any time. If you wanted to just finish one area at a time, you can just find a joint area. For instance, if you do the mustache, the outside line would be the joint. It is similar with the hat, if you just wanted to finish the hat, the bottom edge of the hat would be the joint. Anything else can be lightly scratched and you can come back to the next day. Just make sure that the mud that's not finished is left in so you can raise it out for the finish coat.

Working on nose.

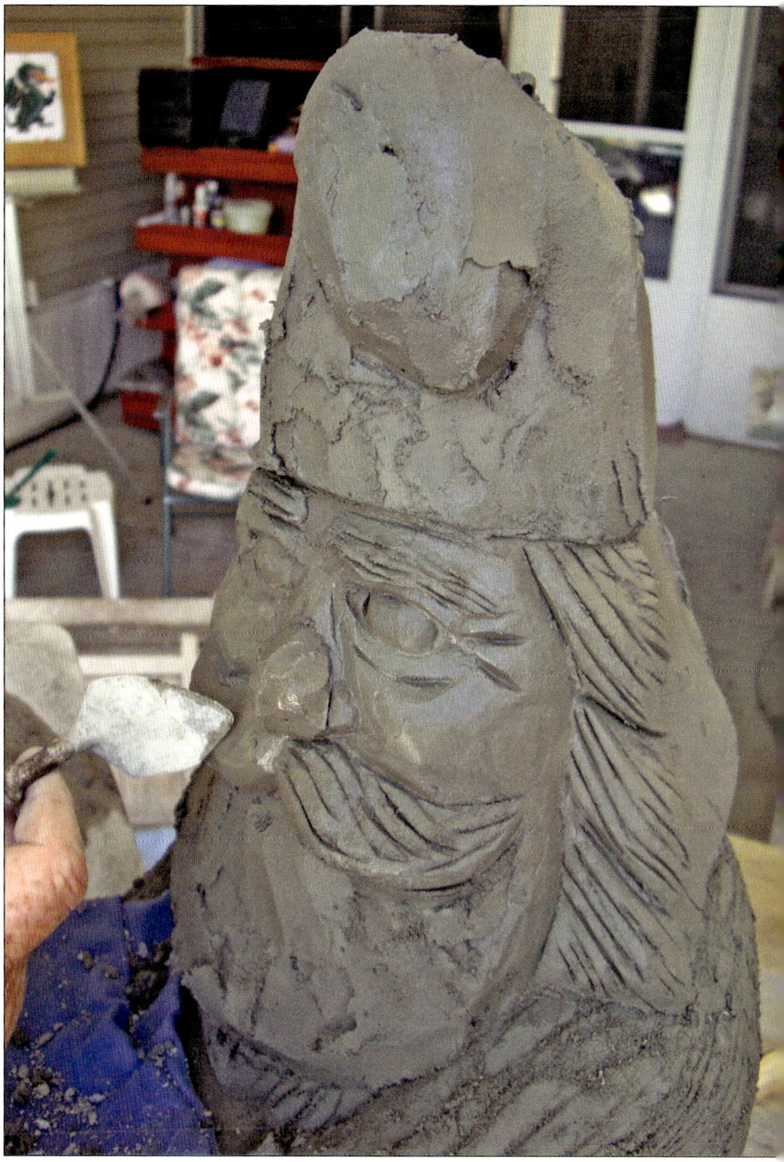

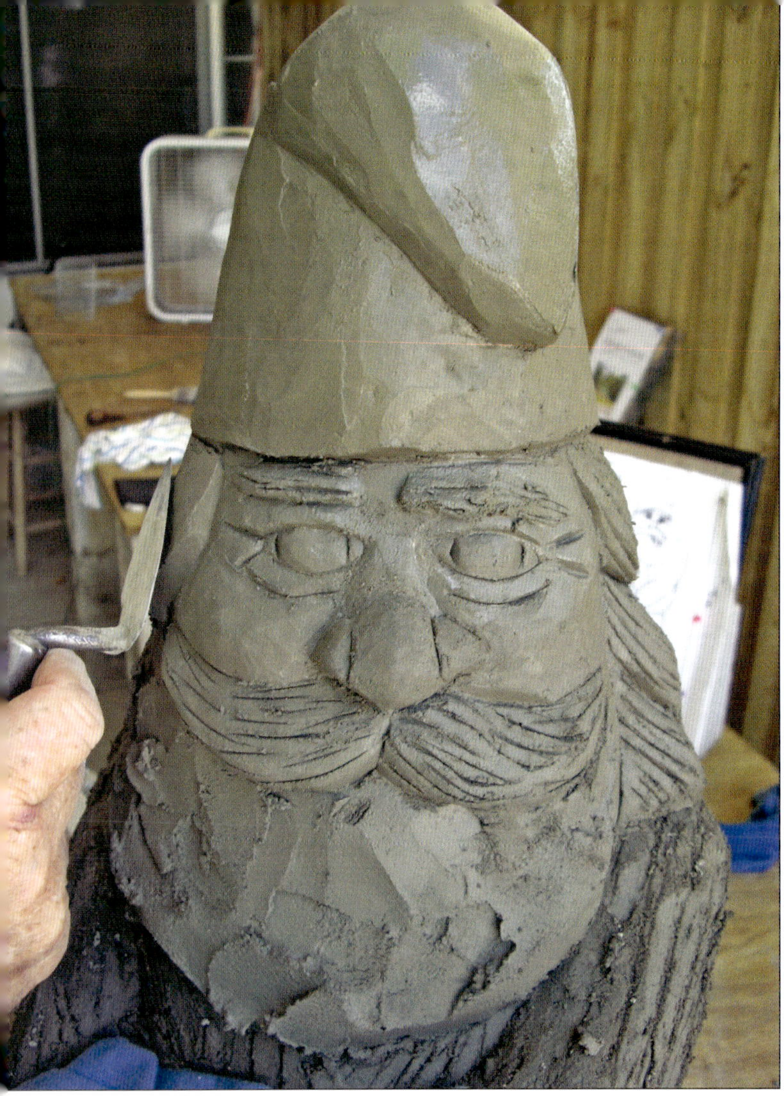

The hat is finished and I am working on beard.

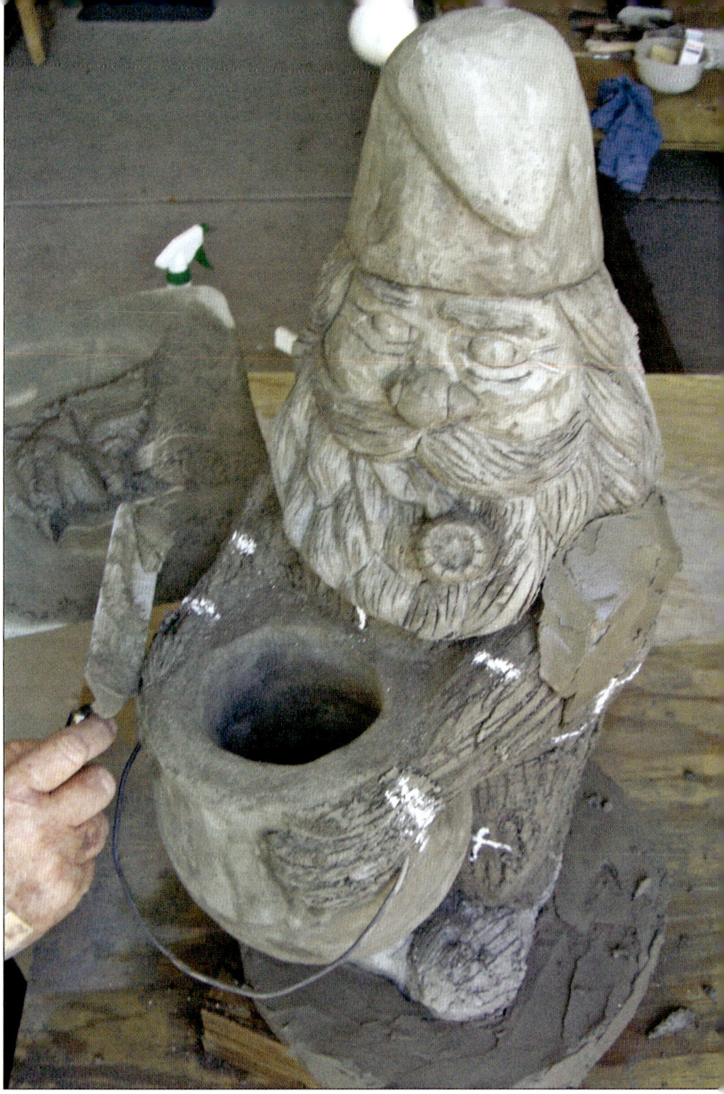

I use chalk on all my sculptures to figure out where I need to build out; example, where the cuff on the shirt will be or where a button will be.

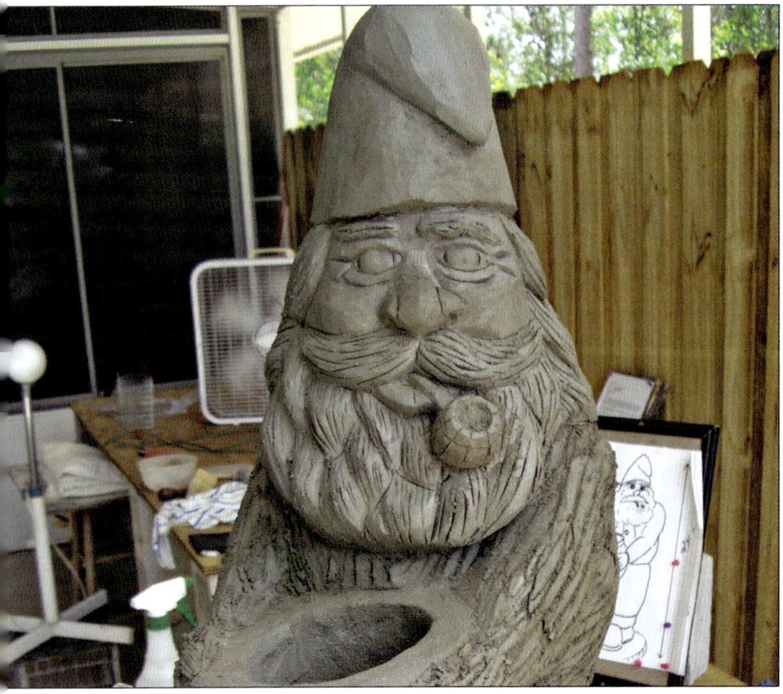

The head is complete.

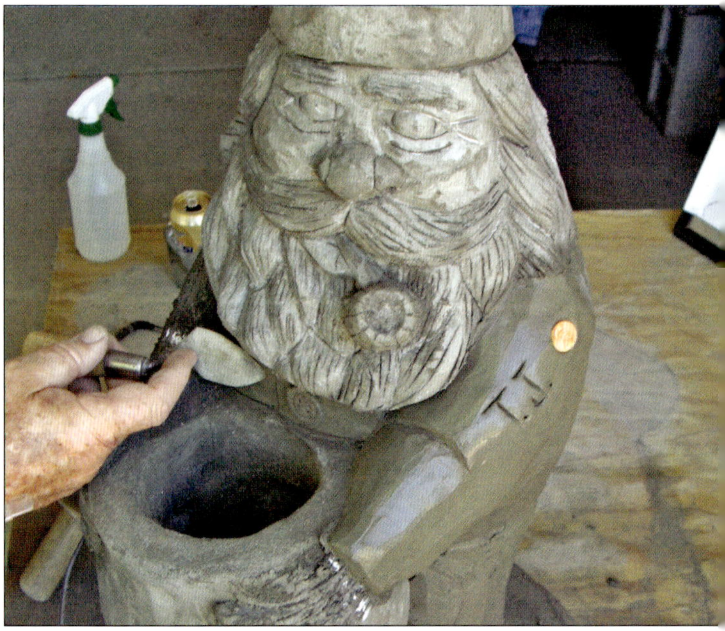

A penny is set in the shoulder for luck and my initials are added.

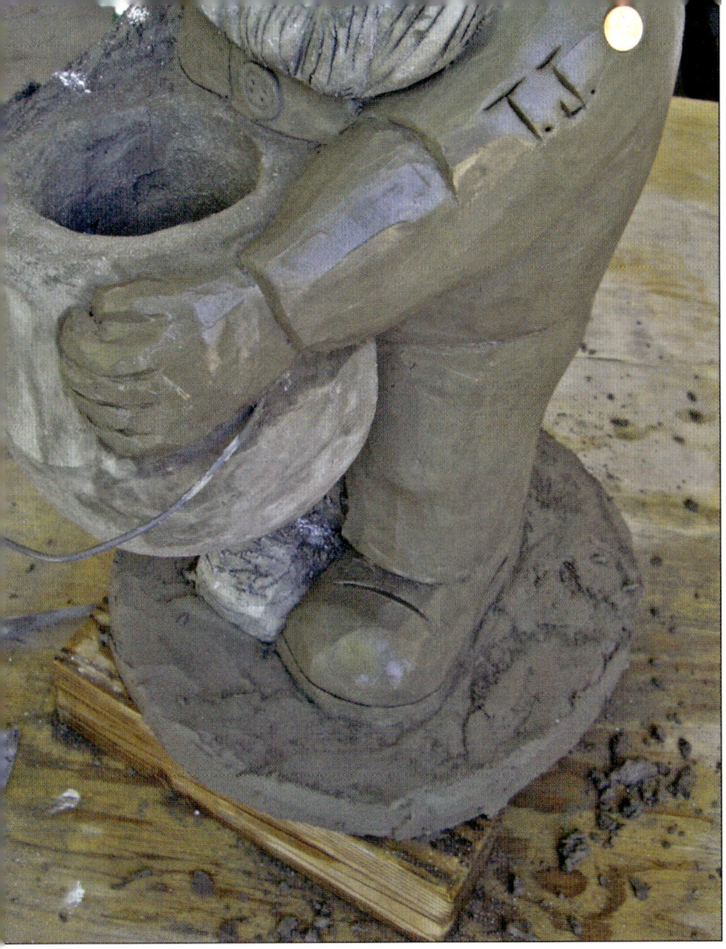

Complete hand, arm, leg, and boot.

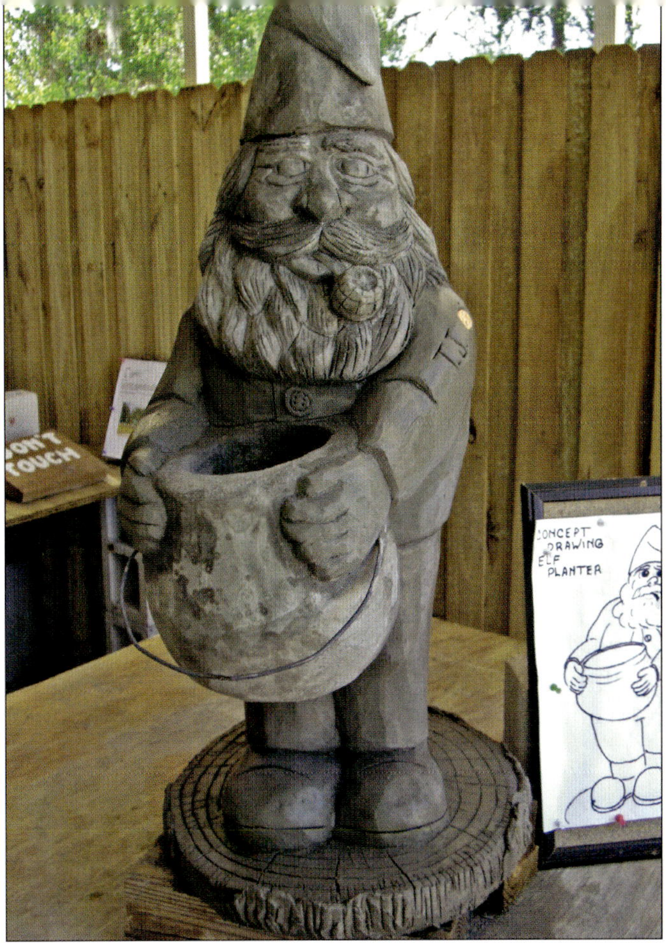

Concrete work is finished.

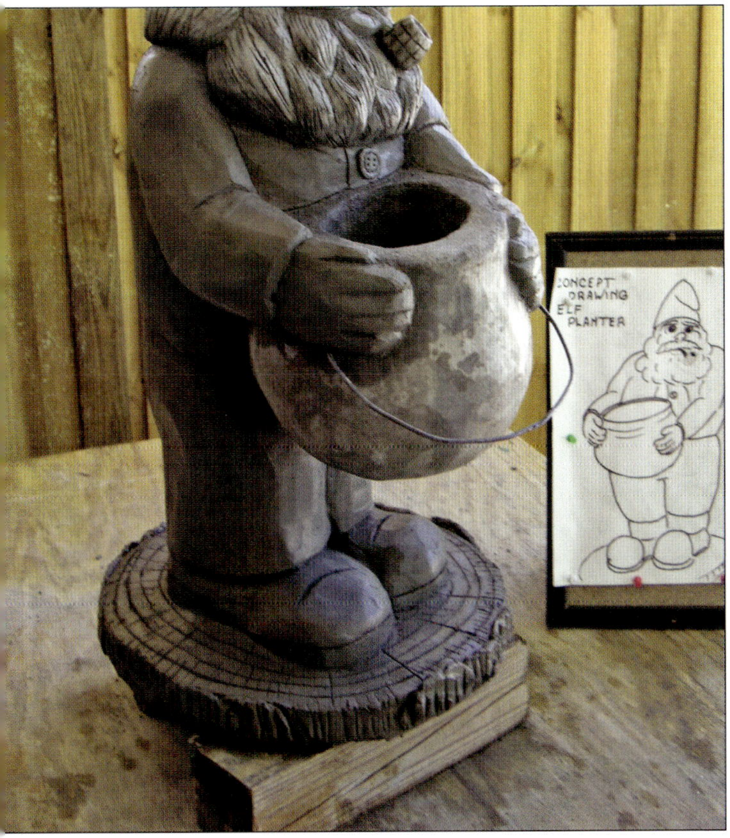

The base of the sculpture has been finished to look like the top of a tree stump.

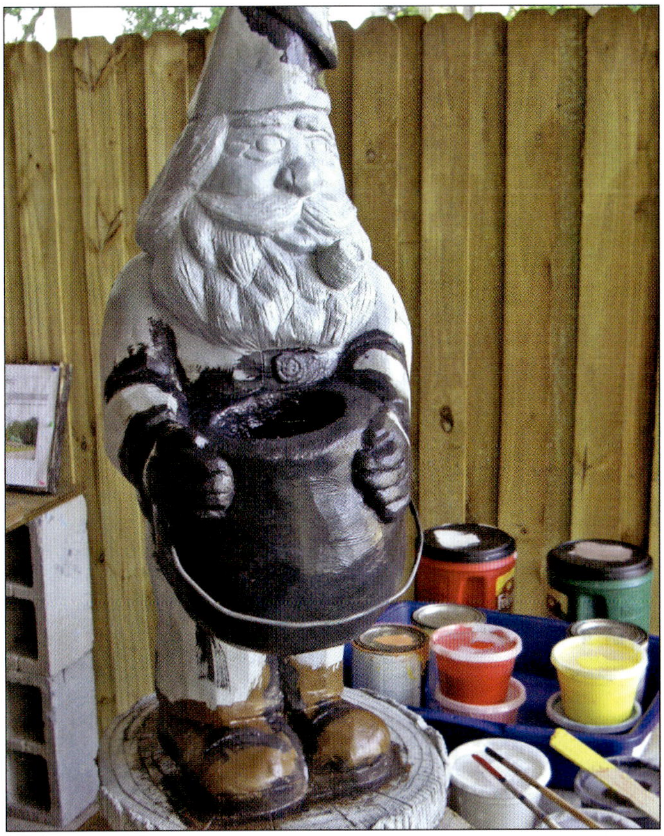

Starting painting.

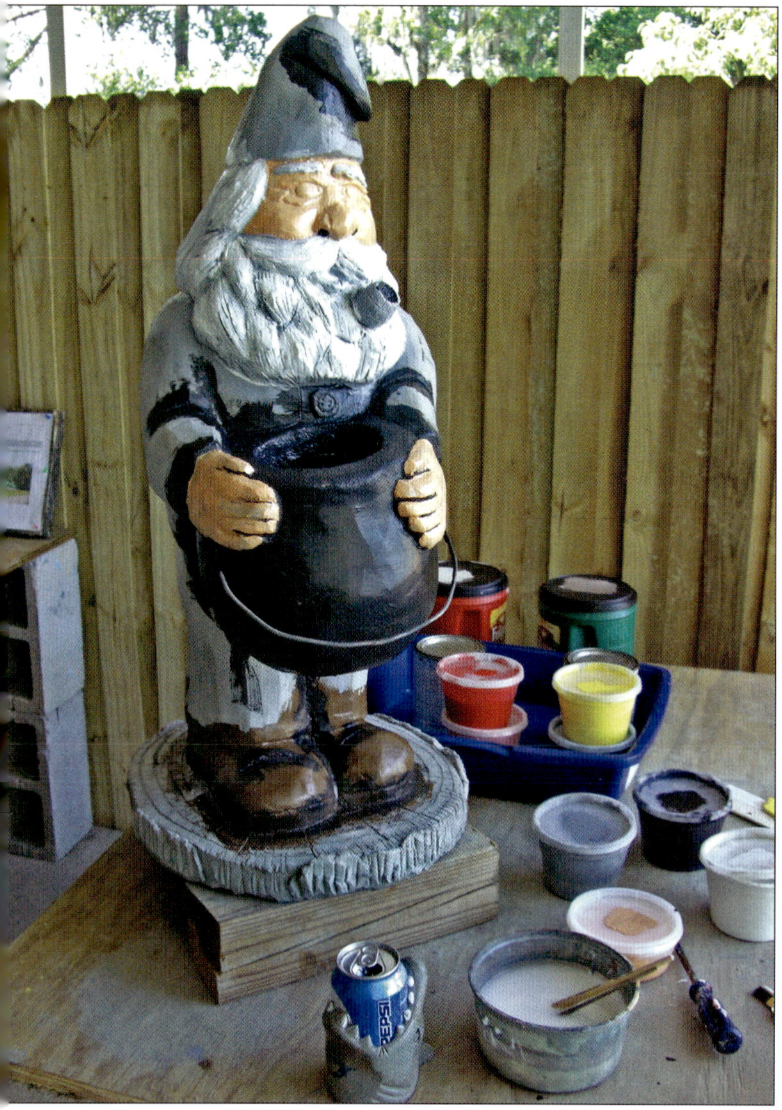

Moving along with painting.

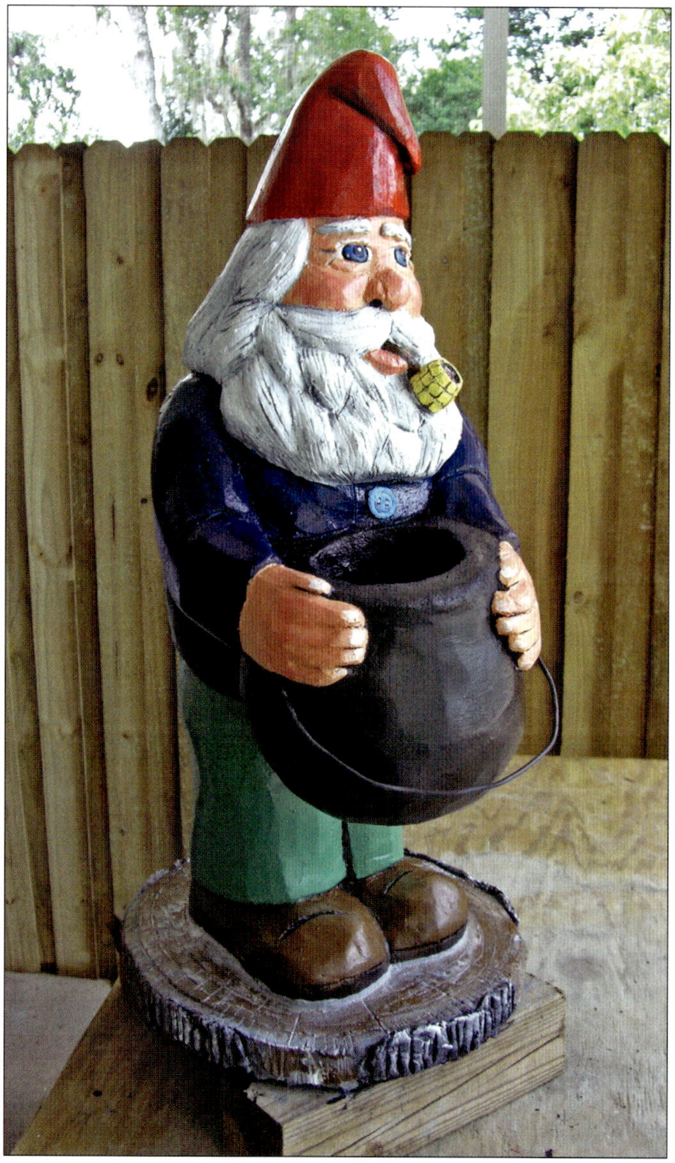

Painting complete.

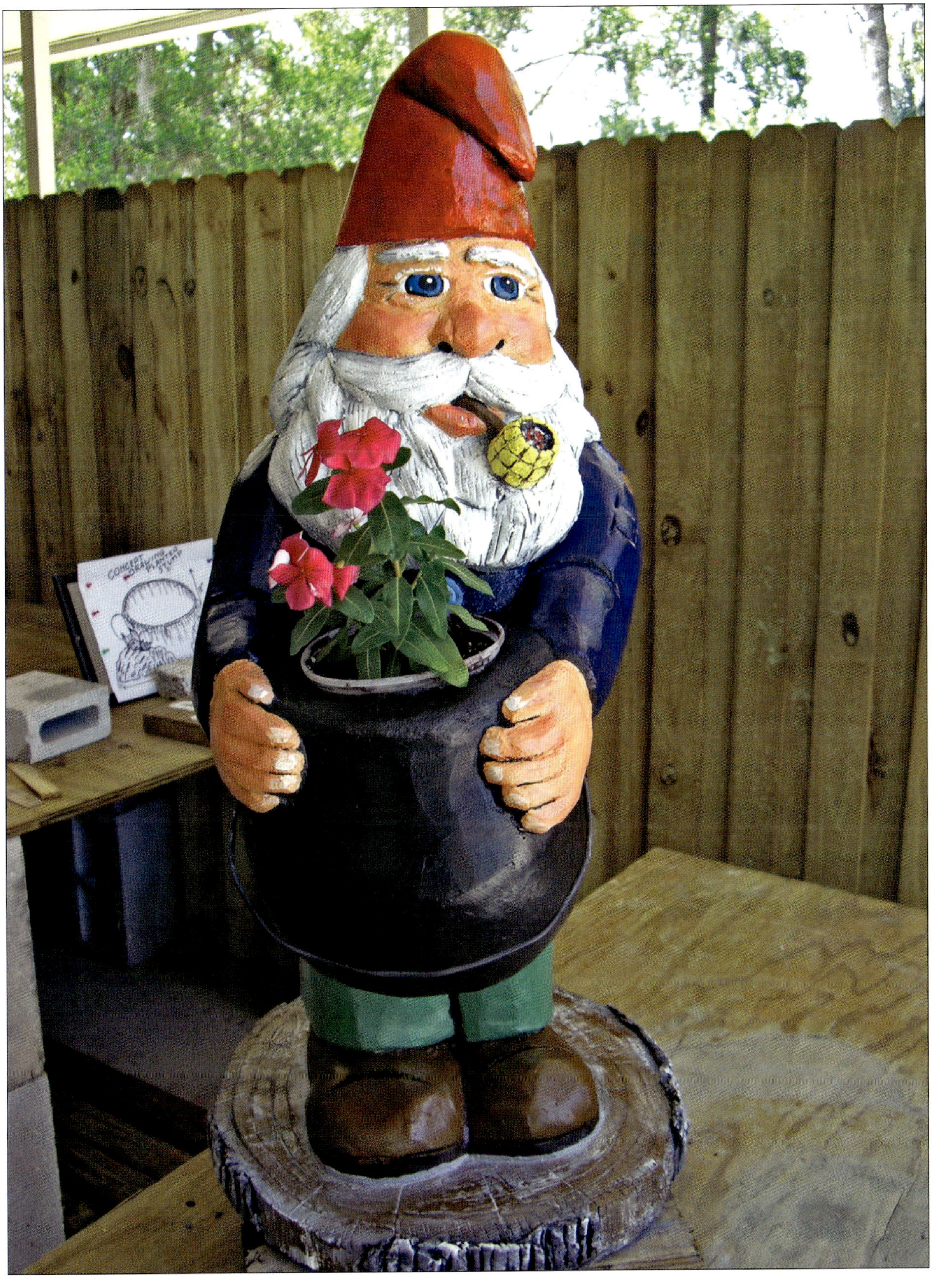

Put flower in pot. My wife likes him so much that she made me give him a name (Gar)!

Tree Stump Planter

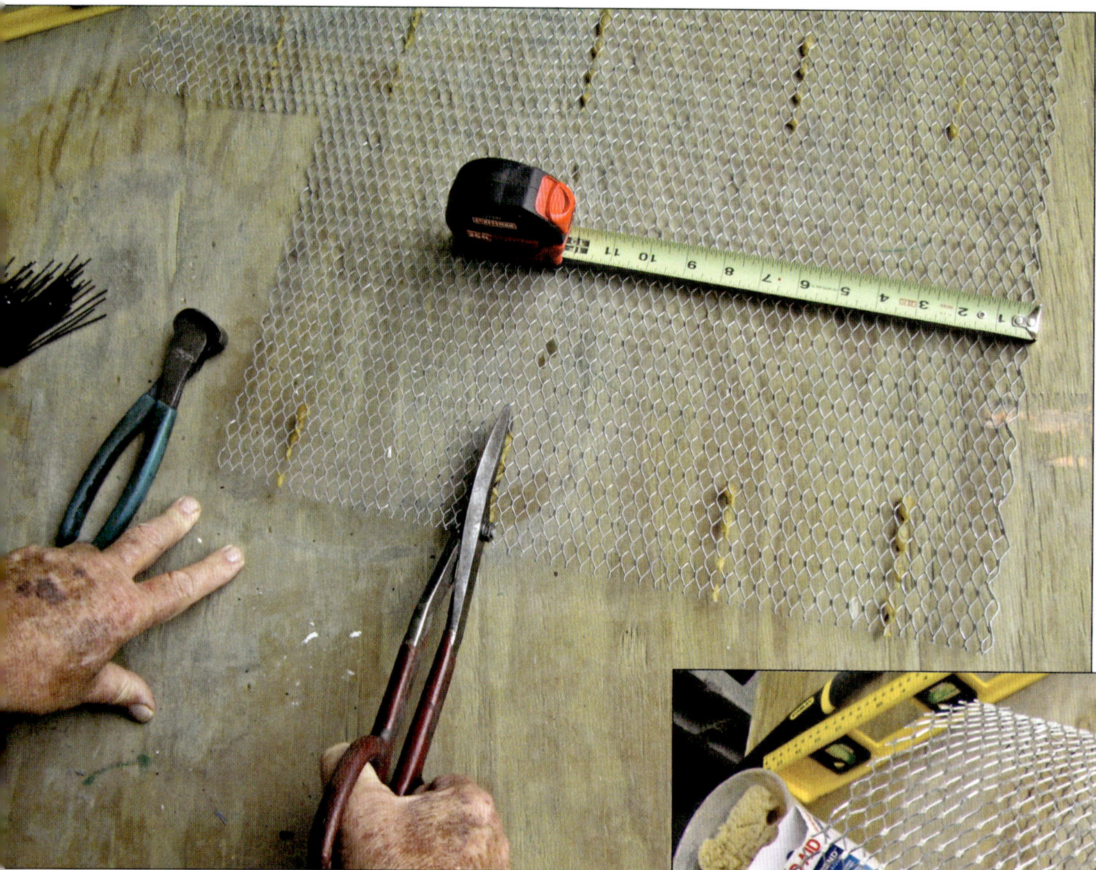

Cutting lathe approximately 12" x 30".

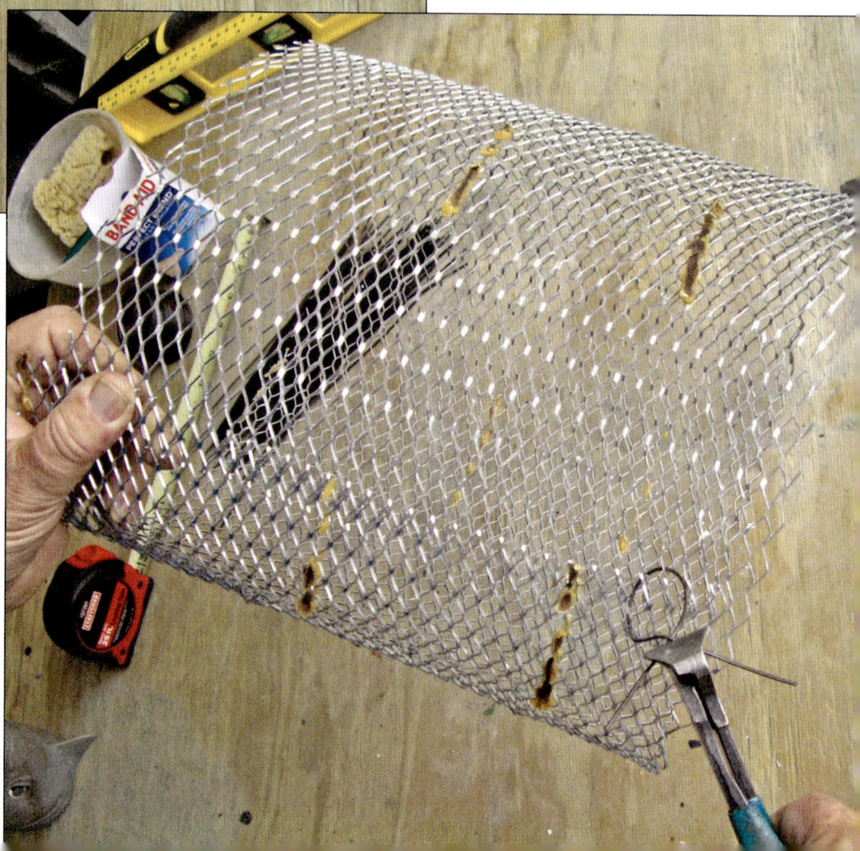

Roll, overlap, and secure with two or three ties.

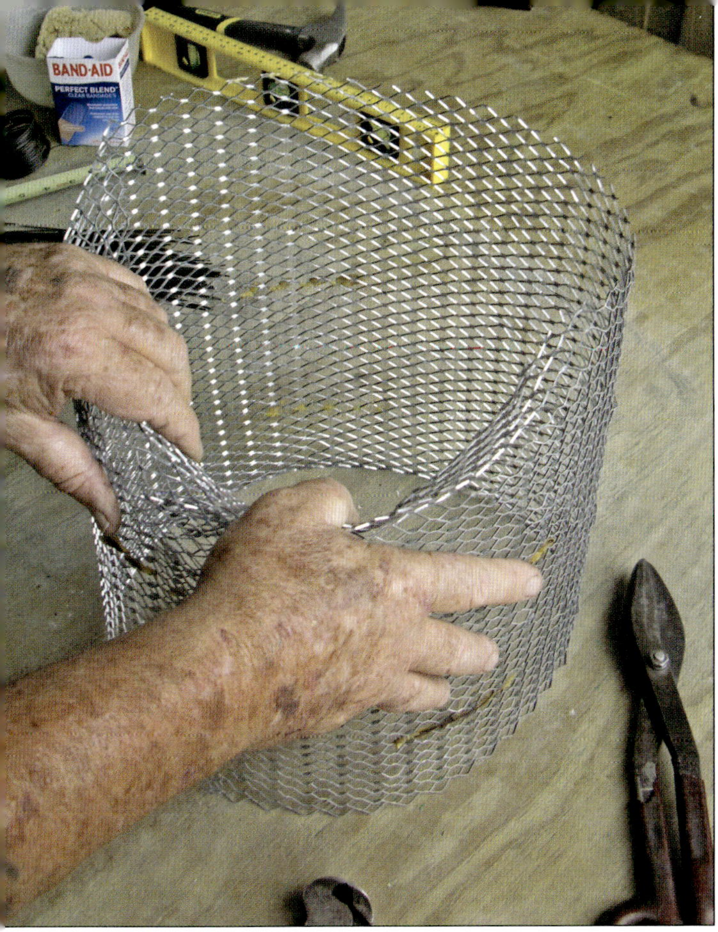

On top of the lathe, using shears, make a cut down 2" to 3", fold sides in and tie. This will be the front of the stump (low side).

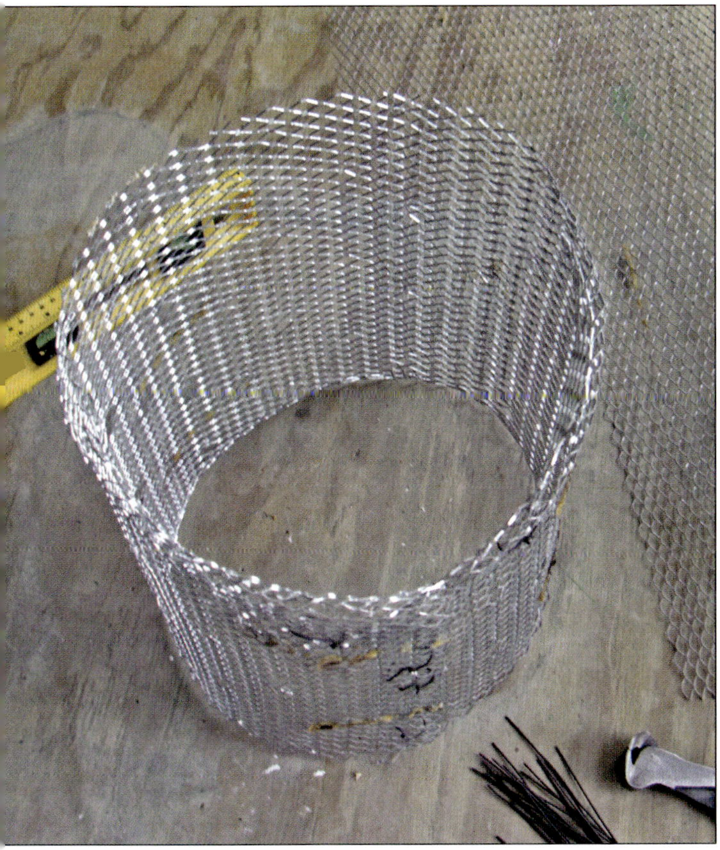

Cut three pieces of lathe approximately 6" x 8".

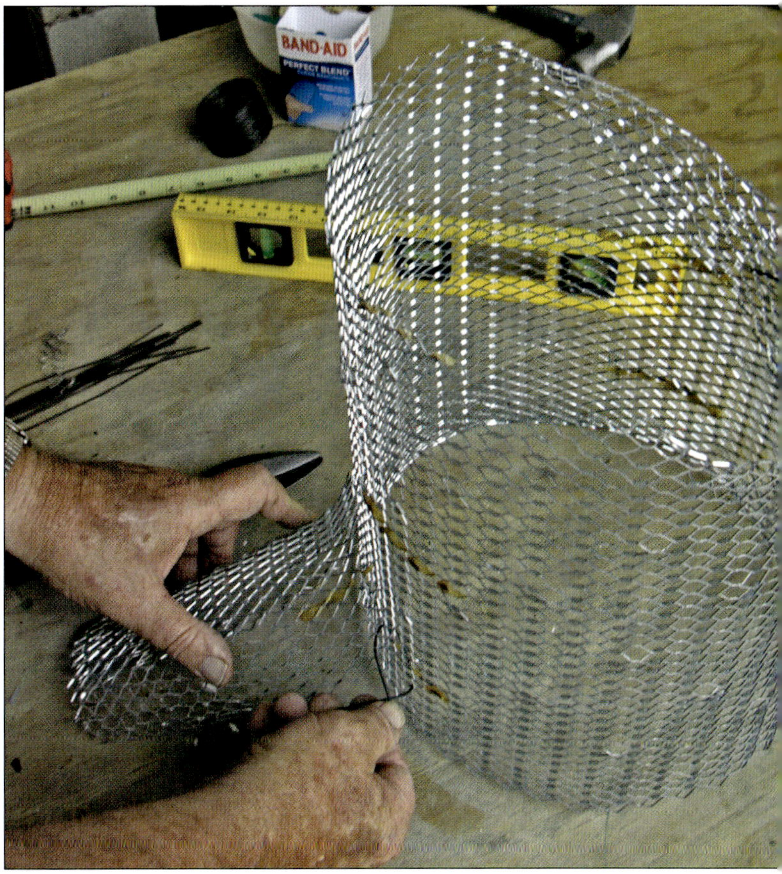

Fold and shape into what you think a root would look like.

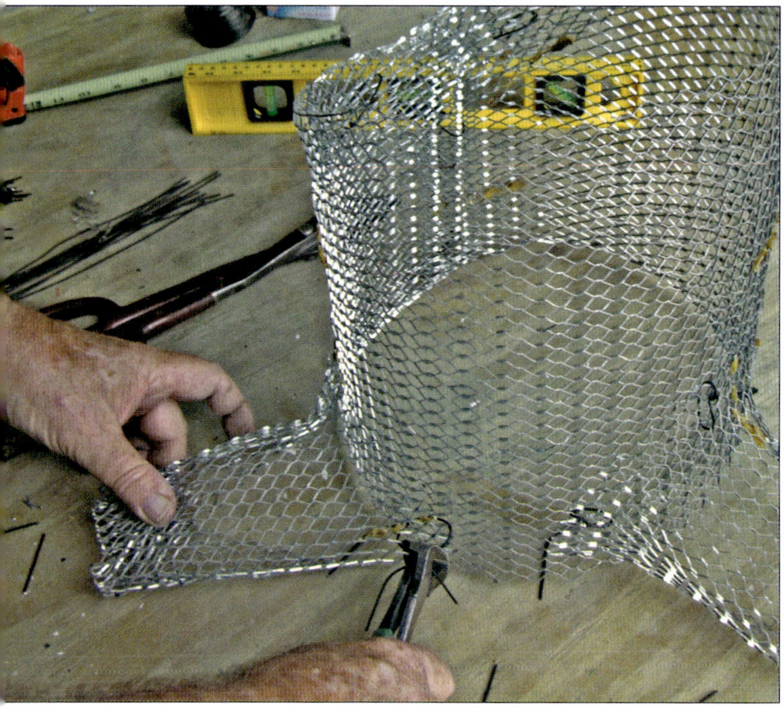

Tie one root on the back of the pot and one on each side.

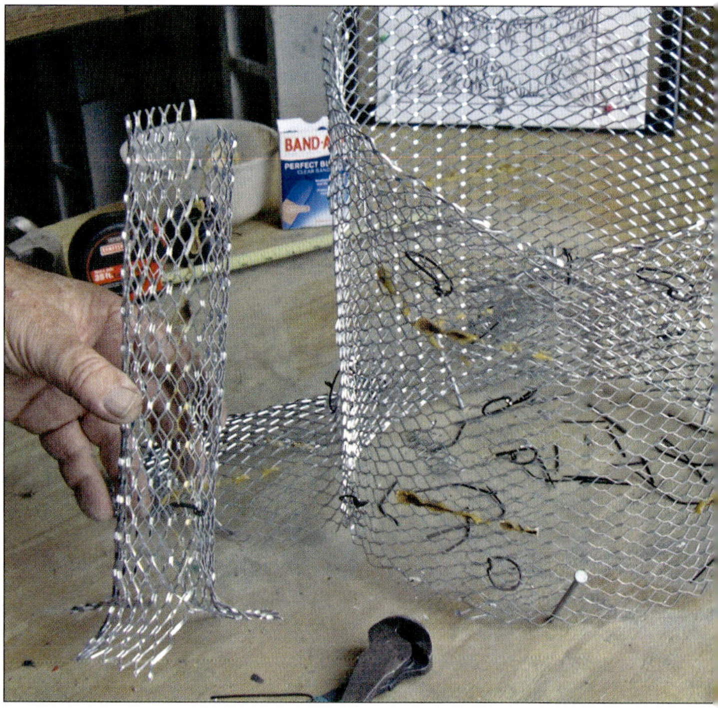

Cut a piece of lathe approximately 12" x 8". Roll and tie. Make cuts with shears 2" on both ends of the mushroom stem. Bend these pieces out.

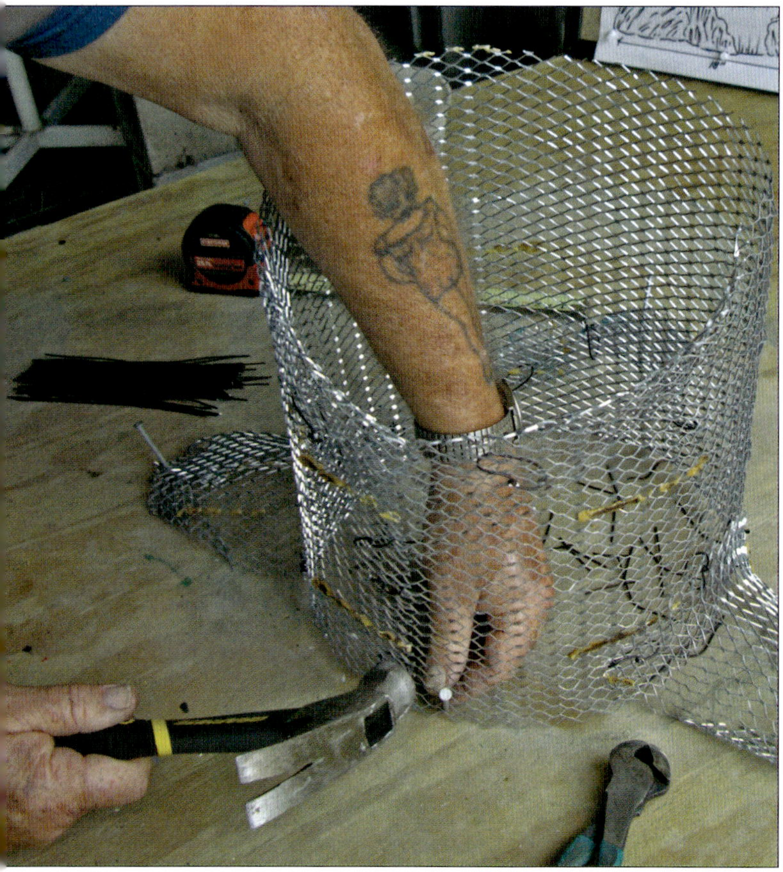

We now need to stabilize the armature by nailing it to the table all the way around.

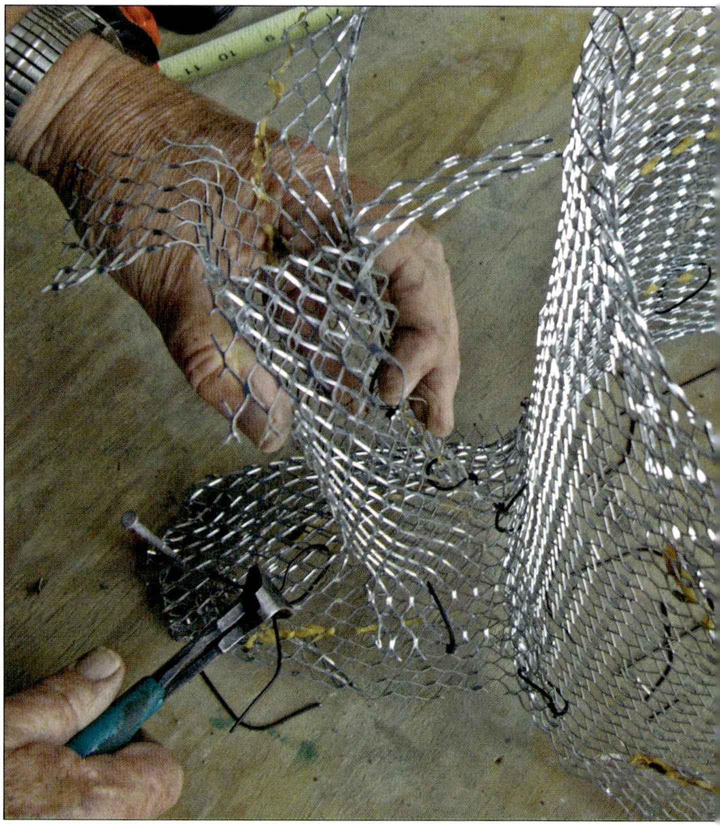

Tie the stem to a root close to the base of the stump.

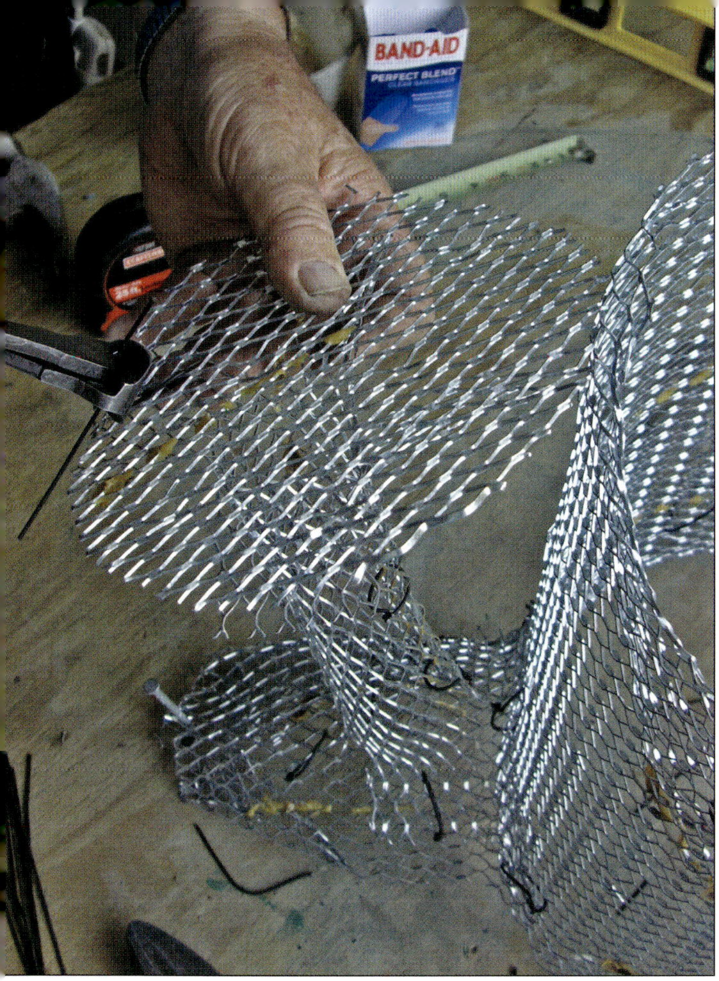

Cut a piece of lathe approximately 8" in diameter for the cap of the mushroom. Tie to the top of the stem using the pieces that are bent over on stem. Make a tie to the top edge of the stump.

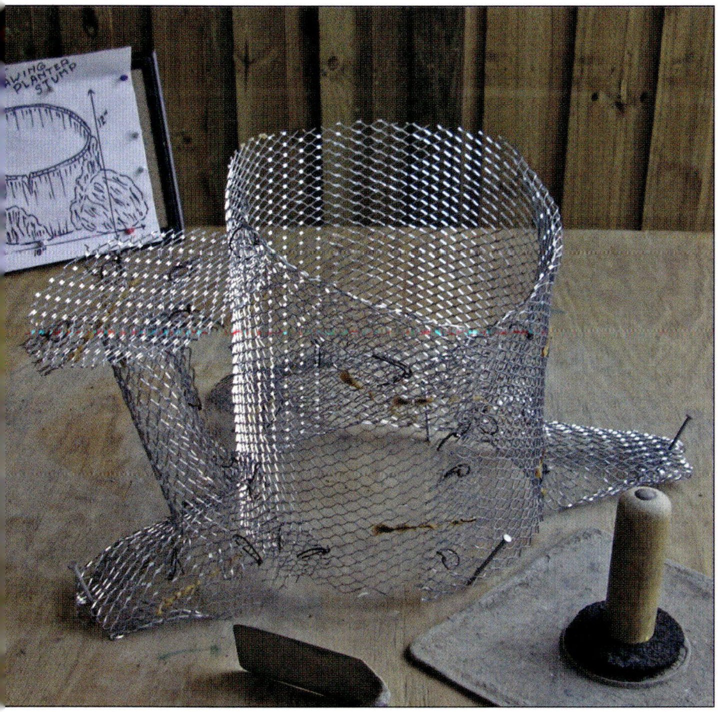

We are ready for mud.

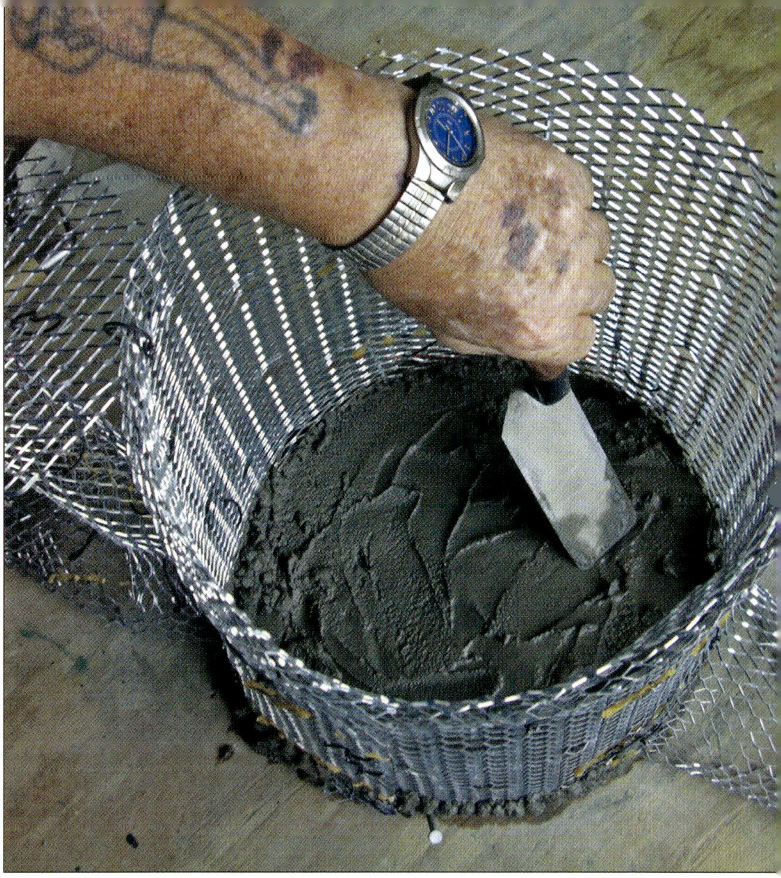

Put a layer of mud about 1" thick inside the stump. You do not need lathe wire inside the stump.

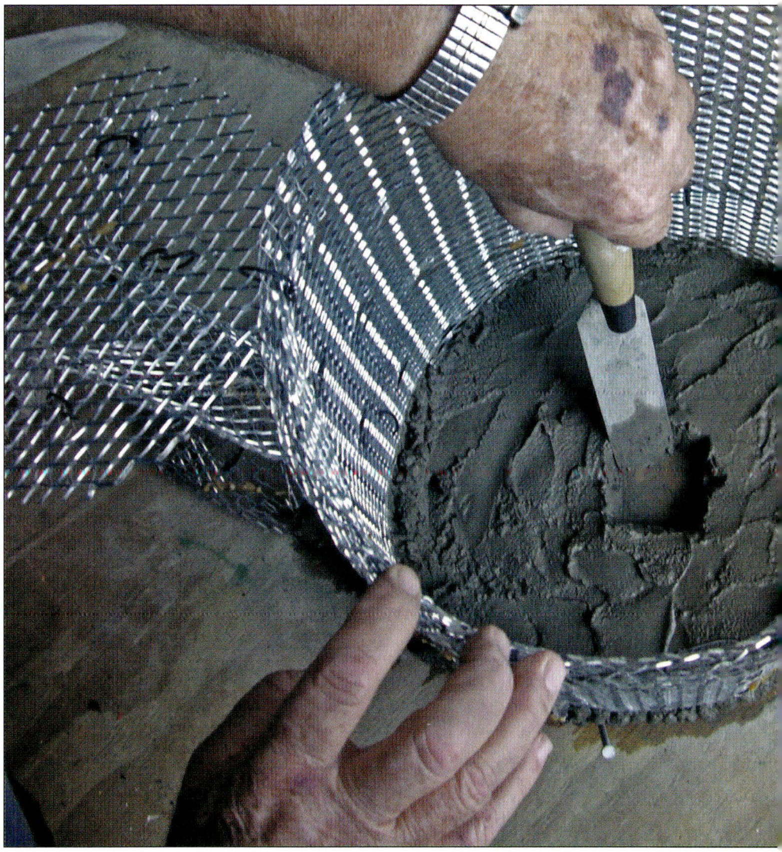

Use a margin trowel to cut a hole in the middle for drainage.

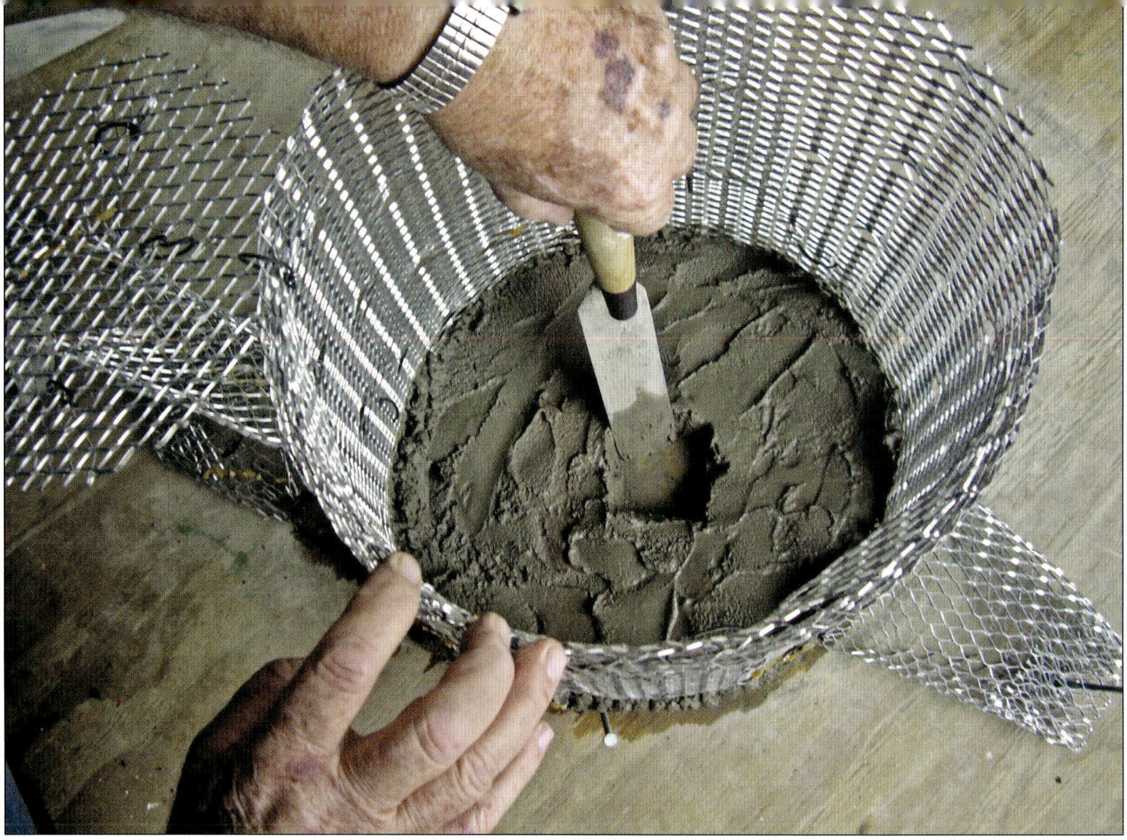

Use a margin trowel to cut a hole in the middle for drainage.

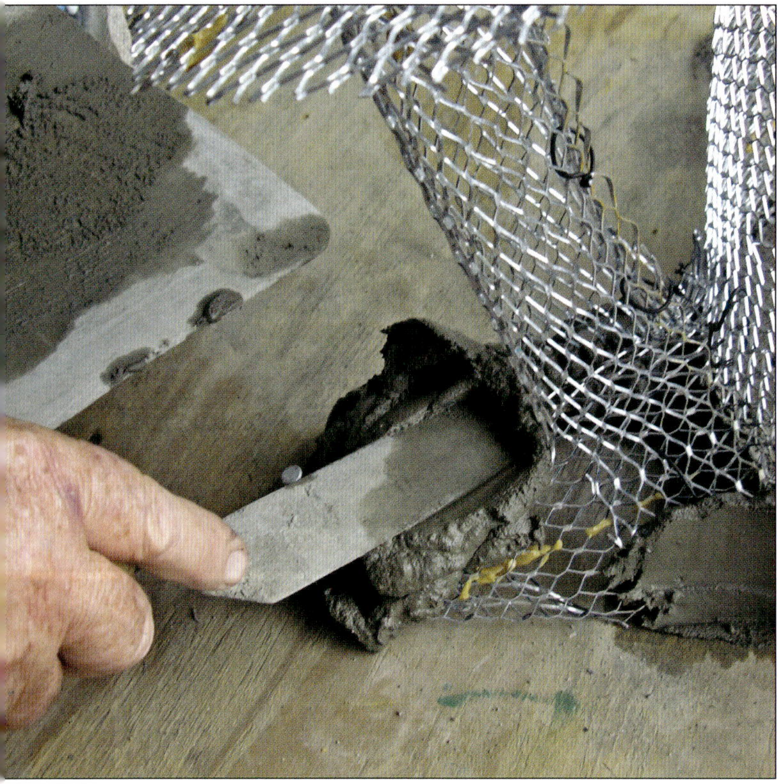

Start mudding the roots. Press mud through the lathe to make each root as solid as possible.

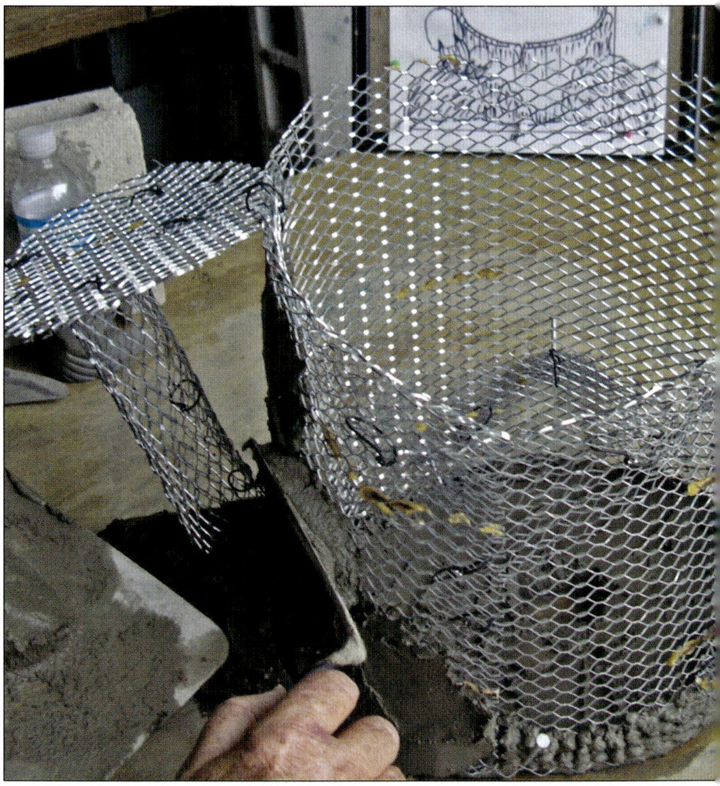

Start mudding the stump.

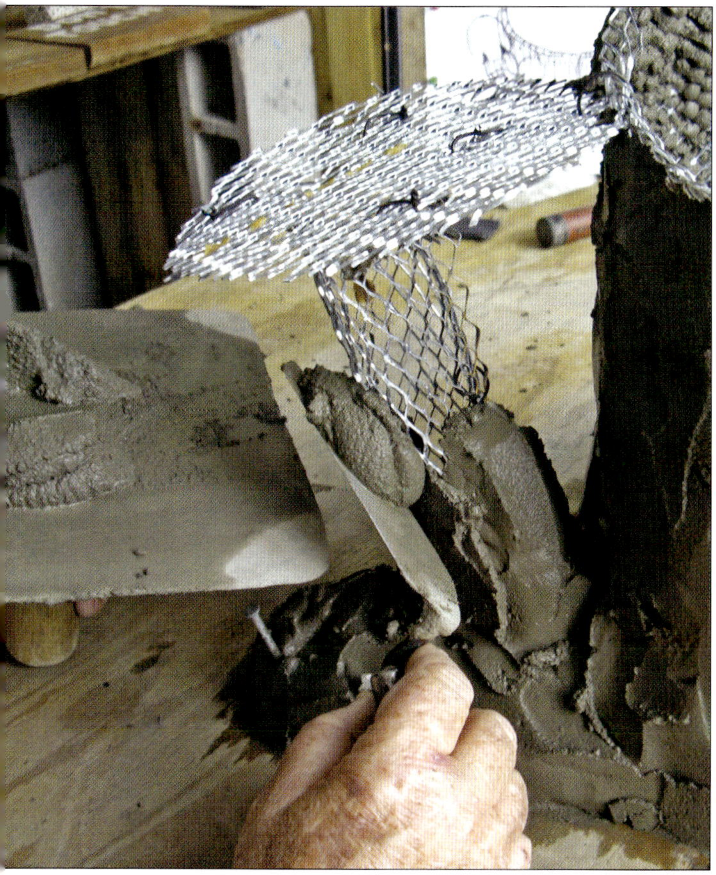

Mudding stump and stem.

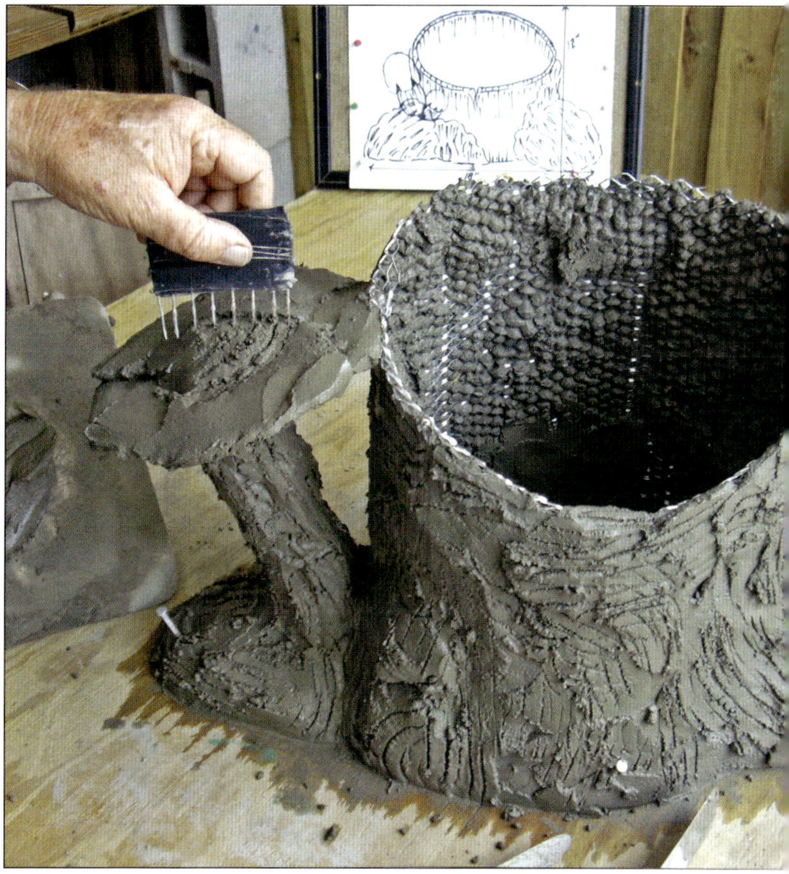

After mud is applied, take your scratch wire and scratch every inch rough.

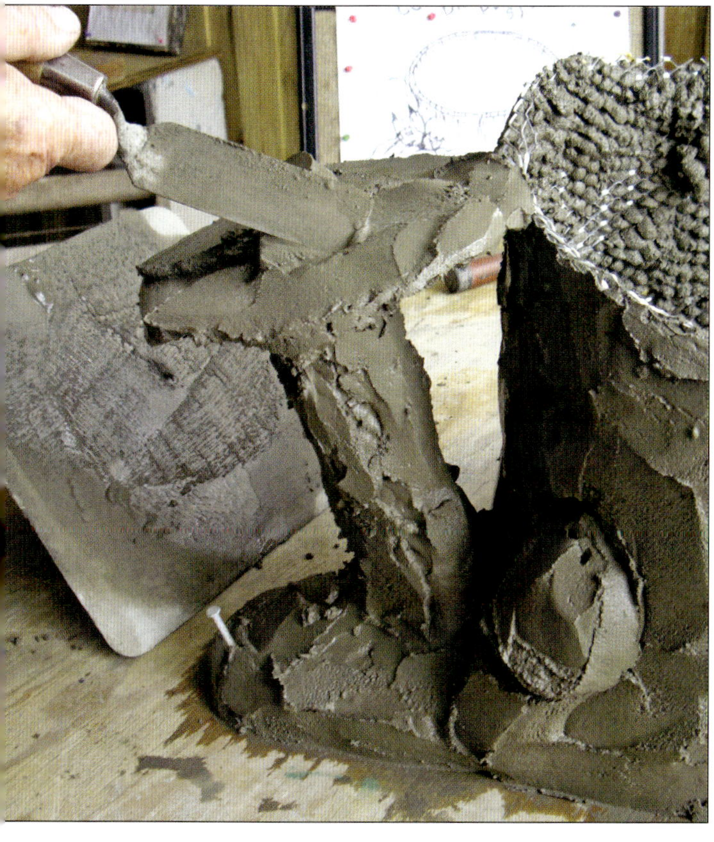

Working on the cap of the mushroom and adding a smaller mushroom (all in mud).

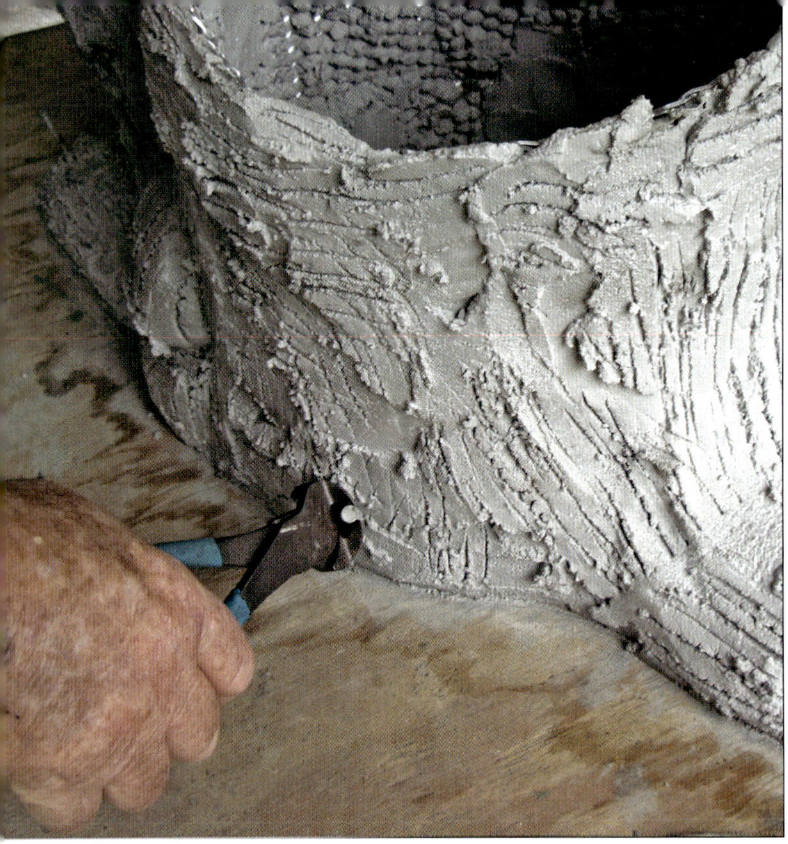

After waiting two days, the stump is nice and hard. Pull out the nails by twisting and pulling up.

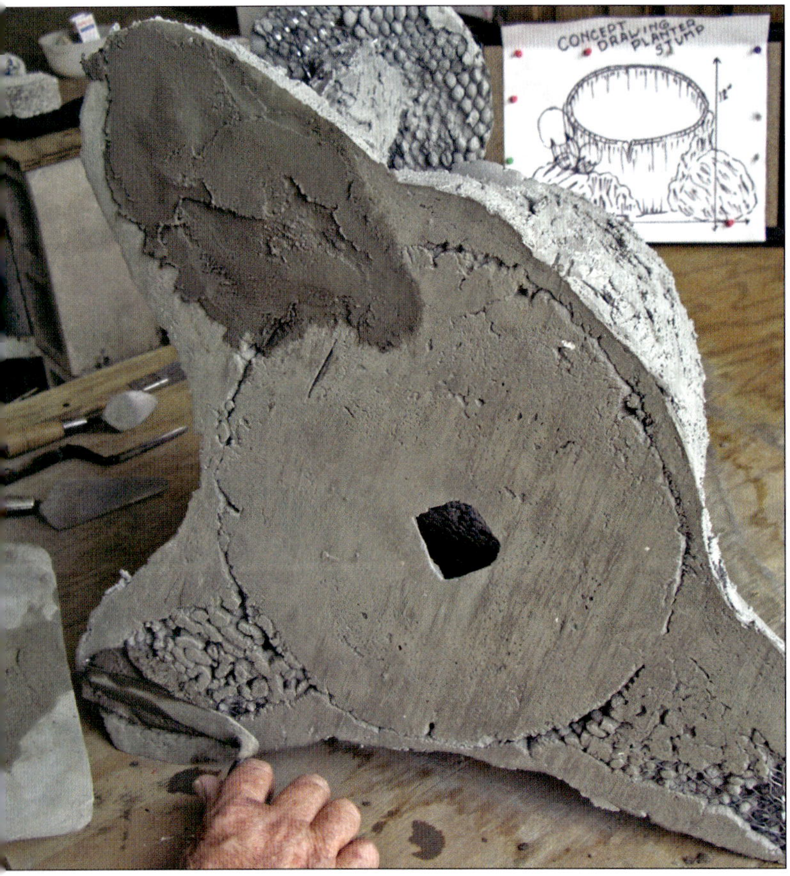

Turn the stump on its side, fill in the areas on the roots that are not full. Notice the drain hole.

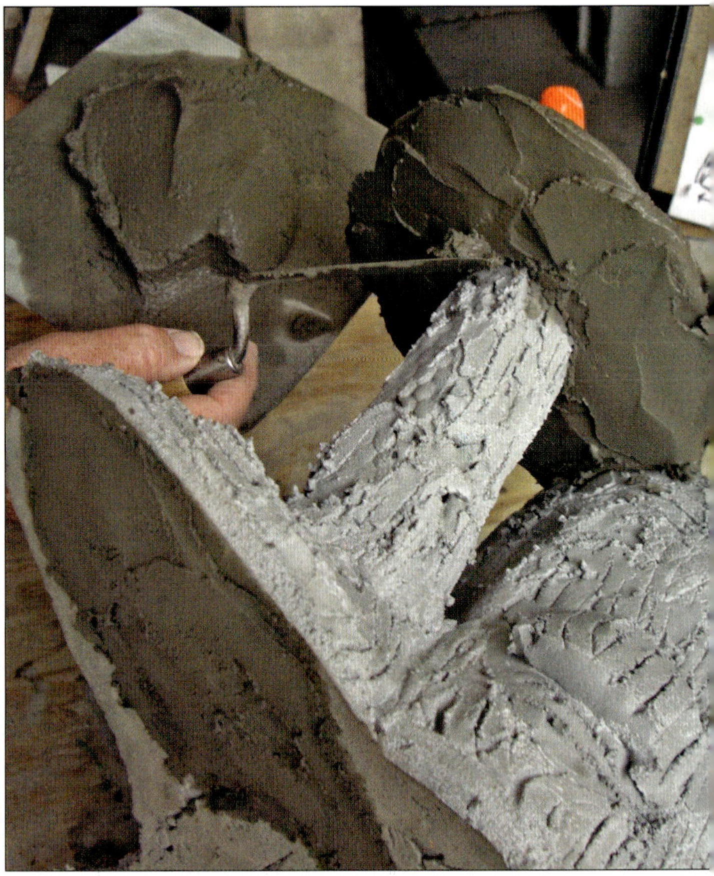

While the sculpture is on its side, start work on the underside of the mushroom cap.

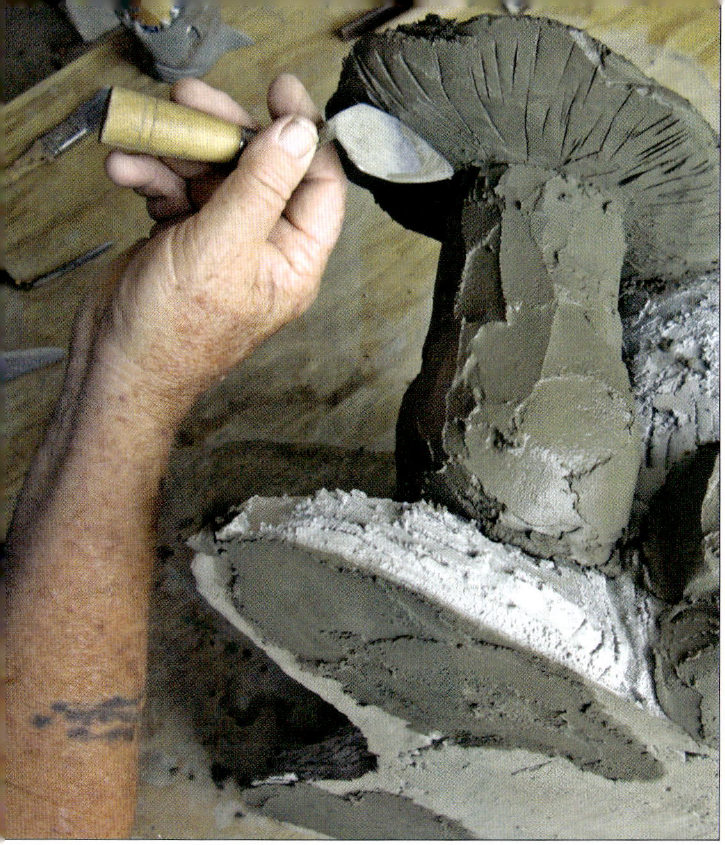

Continue working on the stem and small mushroom.

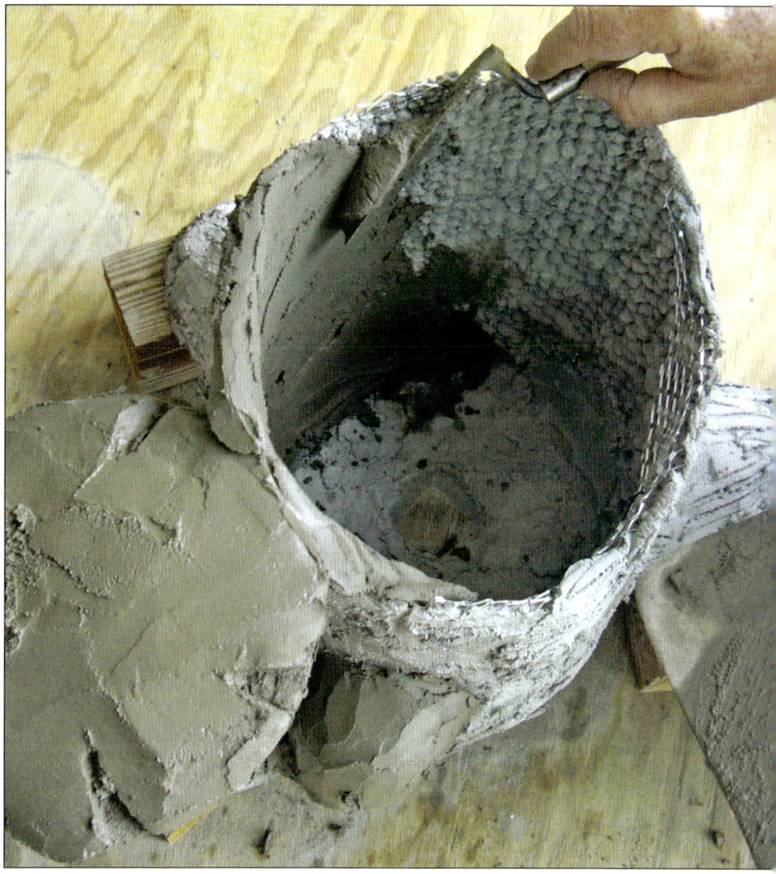

Turn the stump to the upright position. Start work on the top of the mushroom and inside the stump.

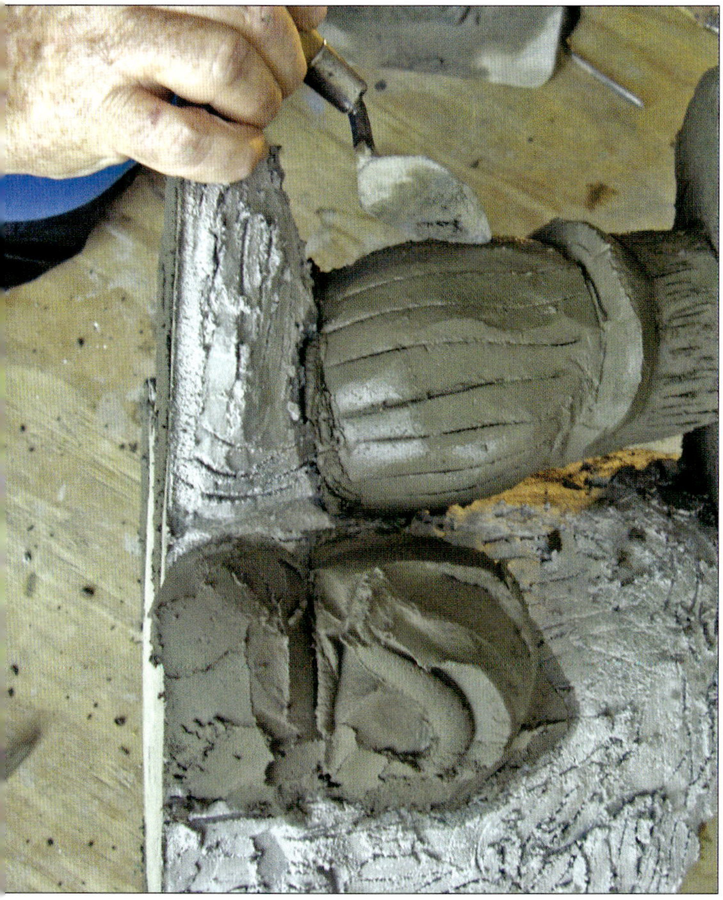

Complete the large mushroom underneath.

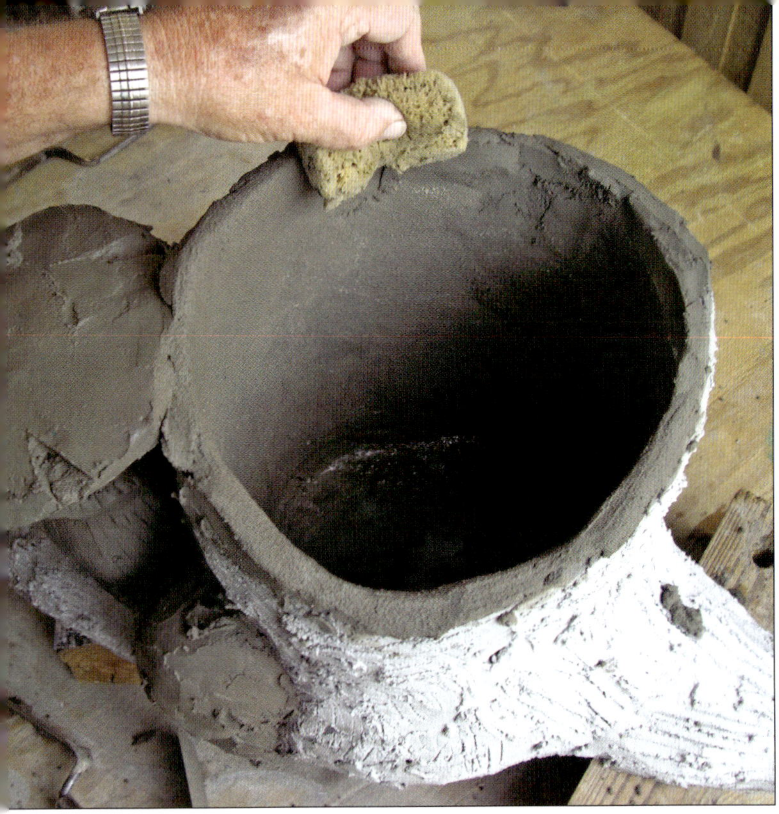

Sponge float the inside of the stump. As I have said in the past, it does not matter if the inside is rough.

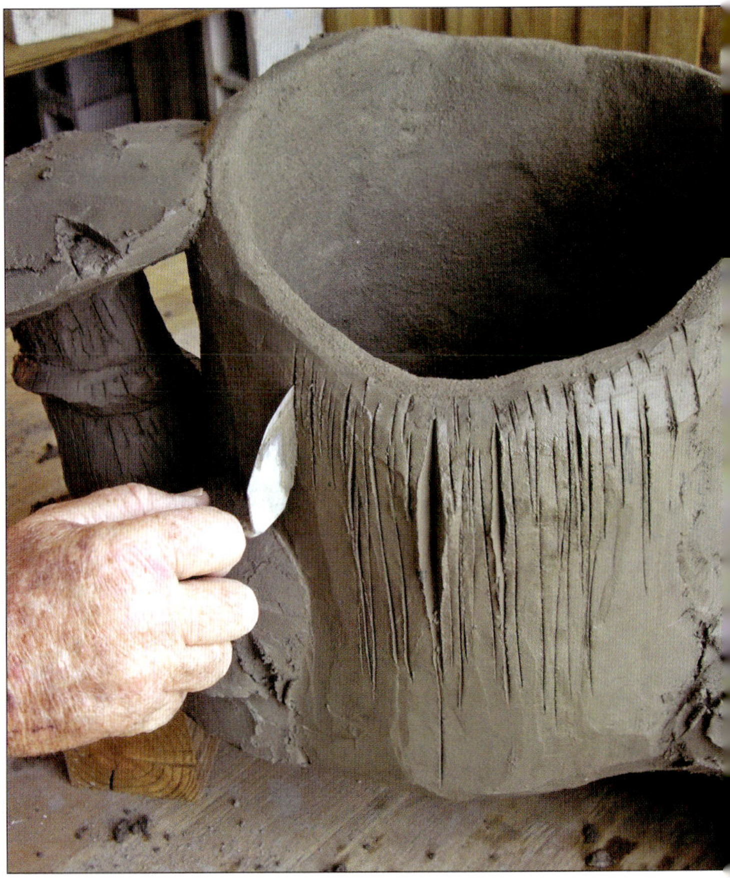

With the point trowel, make cuts and detail around the edges of the stump.

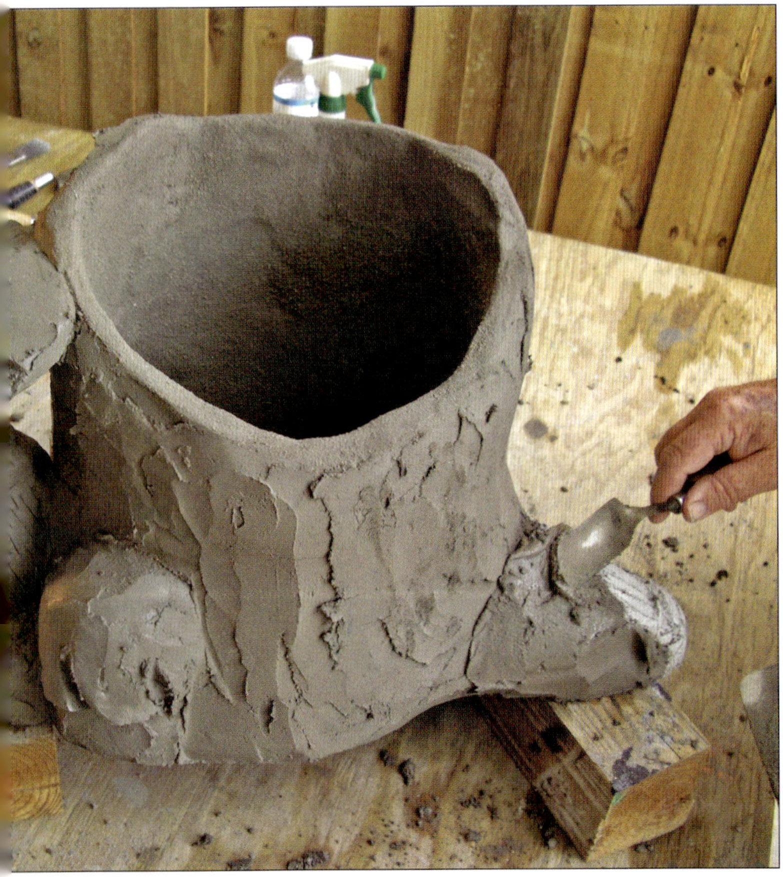

Building out around the stump.

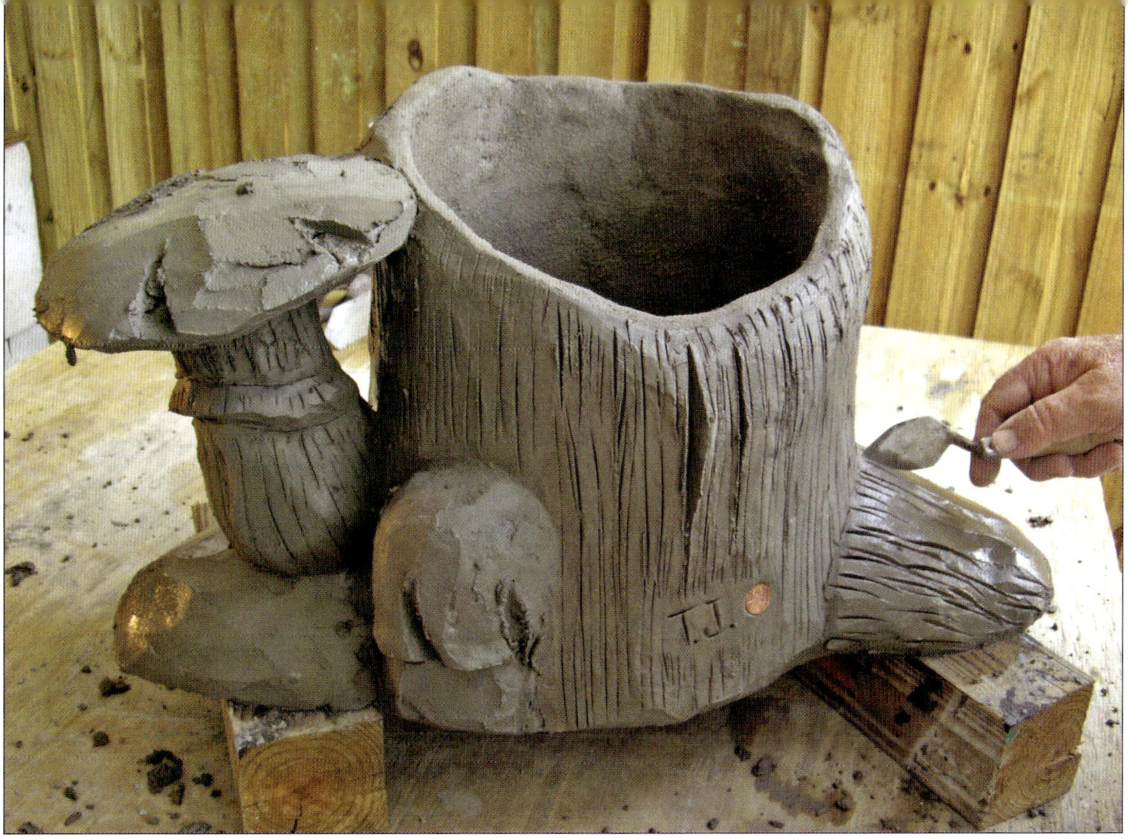

At this time, you can go back and forth, raise out where it is needed on the roots and small mushroom, giving time for thick mud to set before adding more on top.

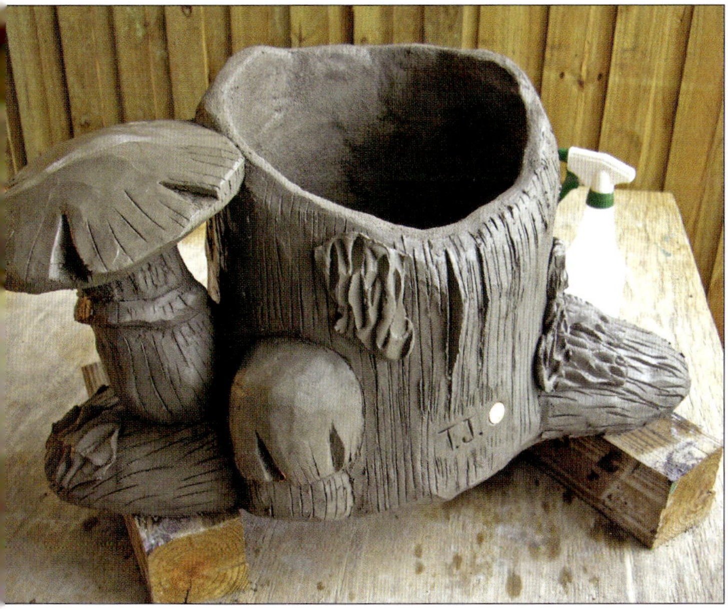

On different areas, add trowels full of mud and then groove that mud to look like bark. Concrete Tree Stump is finished.

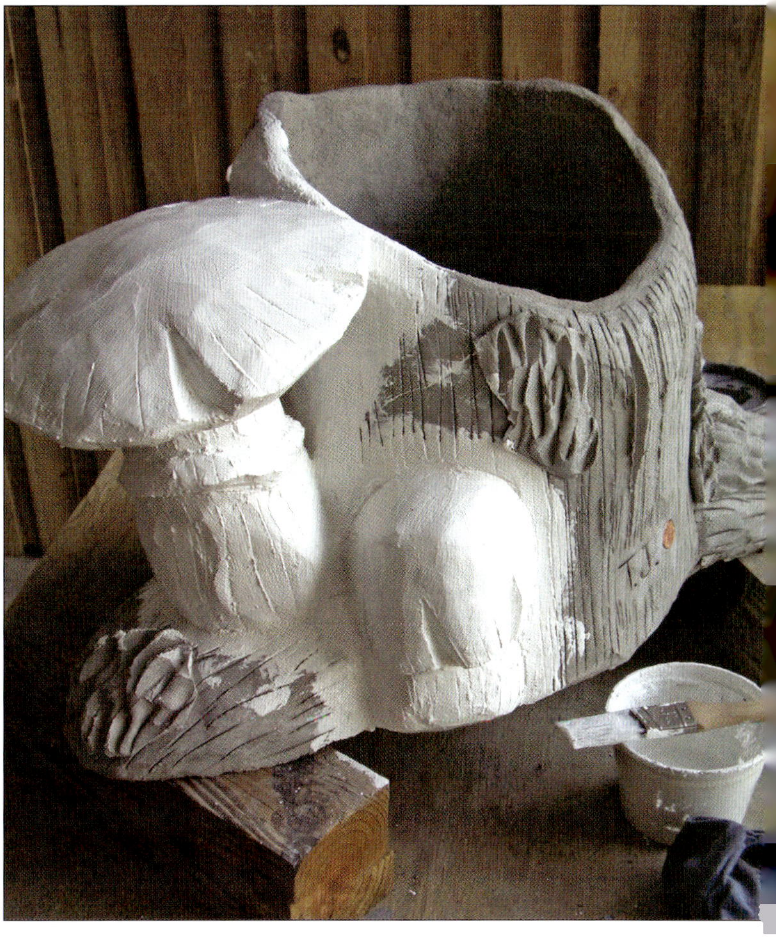

Beginning painting. I started with white on the mushrooms.

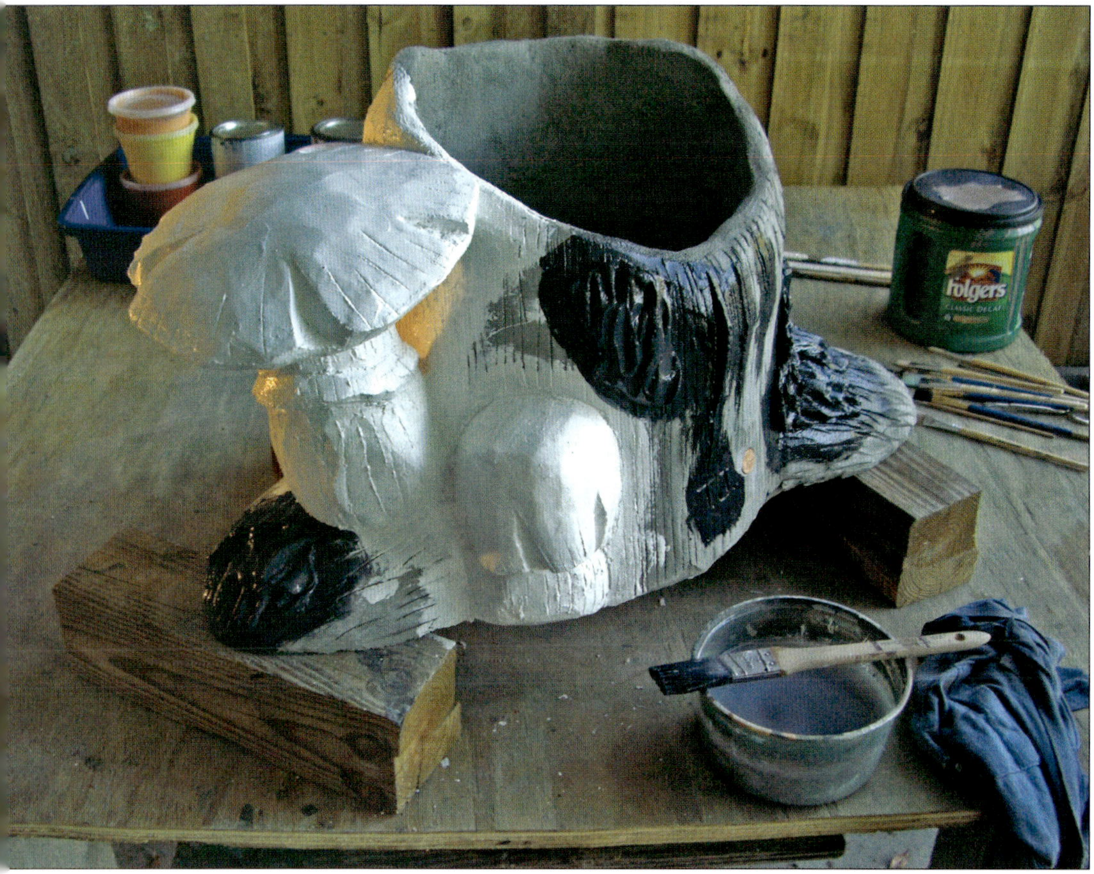

Painted the bark black. Later, grey will be brushed over black.

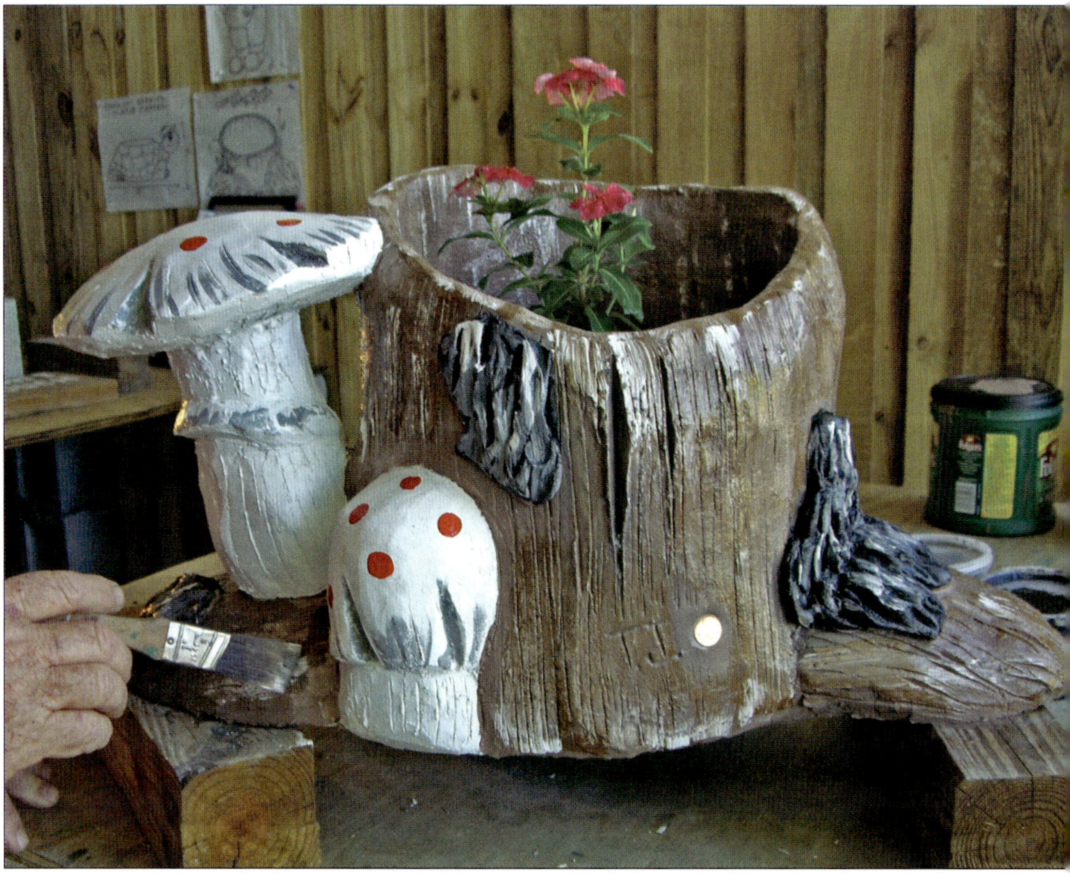

Used medium brown on the stump. Added other colors. Tree stump is complete.

Dolphin Fountain

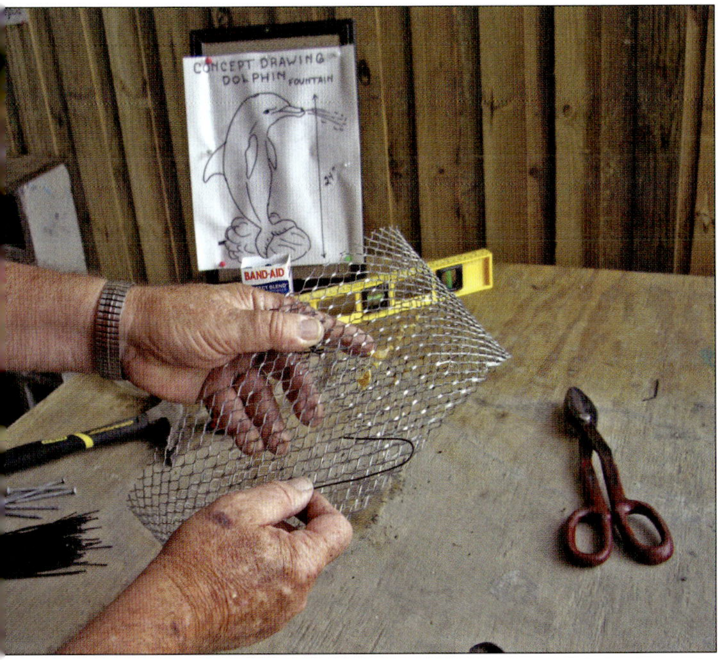

Cut strip of lathe wire approximately 5" wide x 32" long. Overlap and tie together.

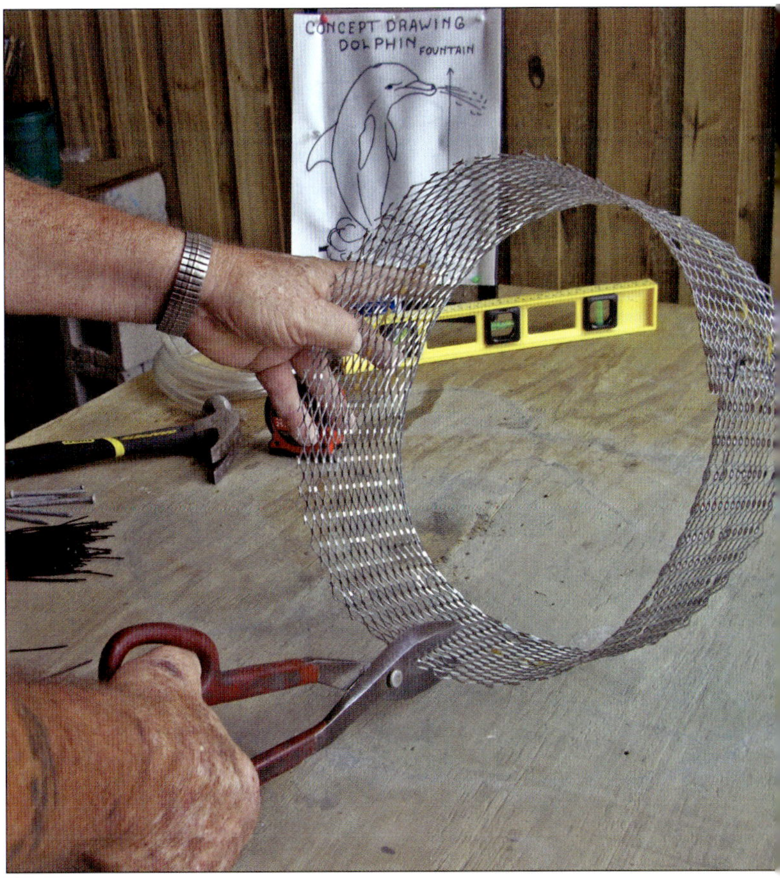

Approximately 1 1/2" cuts, with shears, all the way around the base wire.

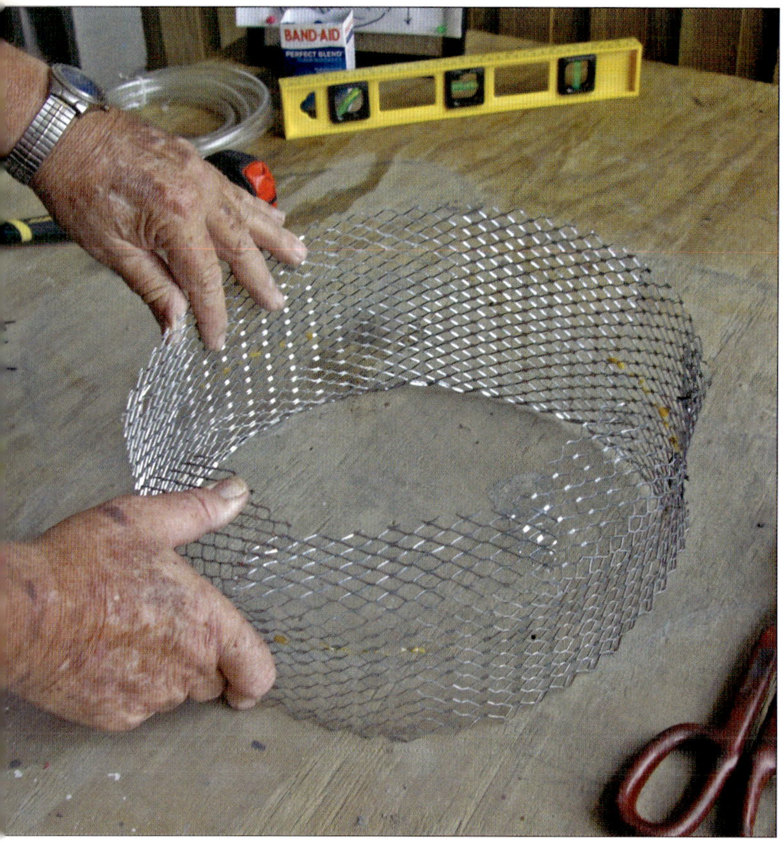

Bend these sections down all the way around.

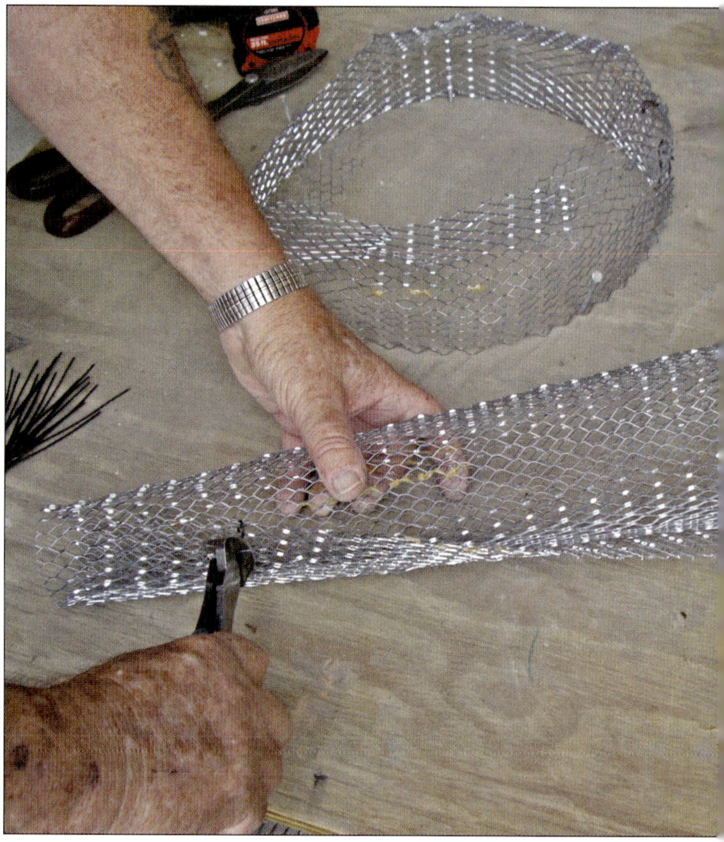

Cut a piece of lathe approximately 12" x 30". Roll into a cone shape. Overlap and tie all the way up.

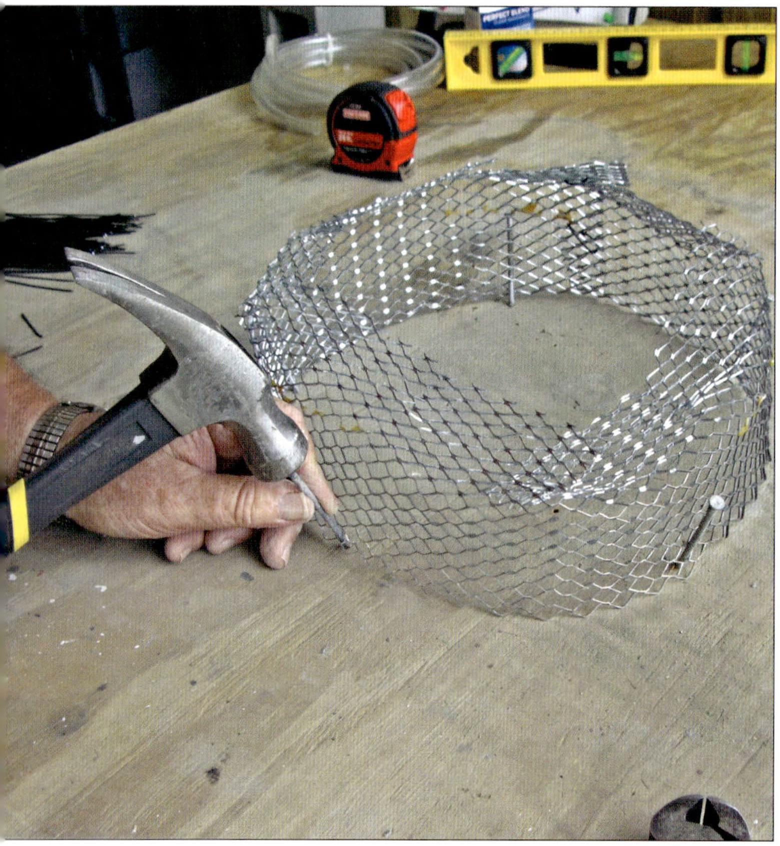

Using four nails, secure lathe to wood.

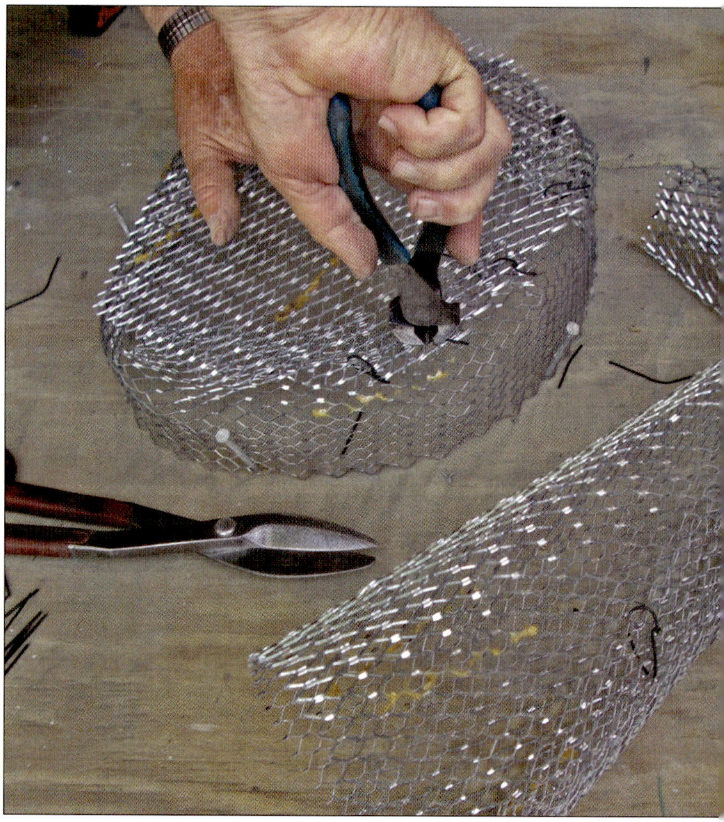

Bend the top part of the cone down approximately 6". Then cut a piece of lathe to fit over the base wire and tie.

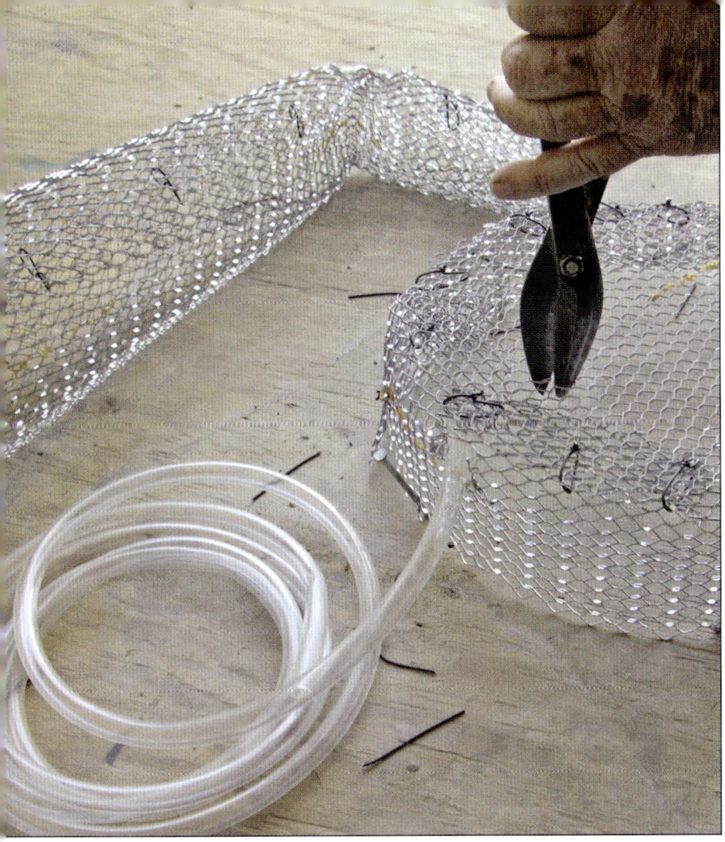

Now, take a piece of 1/2" rubber tubing 4' long. You need it long because you're not exactly sure how tall the sculpture will come out. Cut two holes, one on the top, and one on the side of the base with your shears.

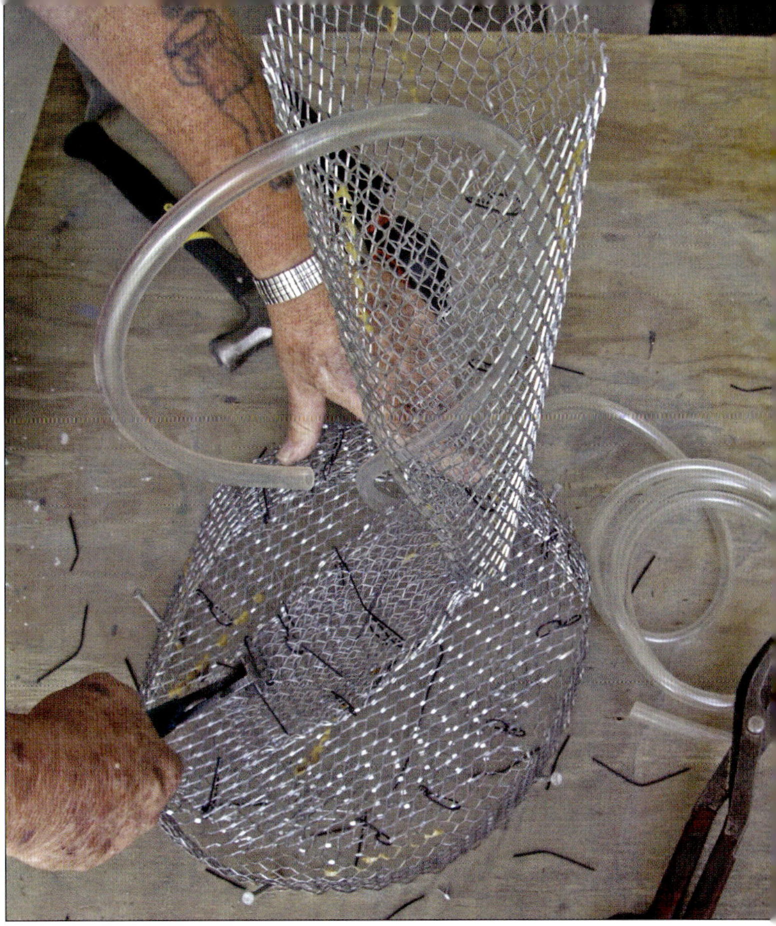

Make a hole in the back of the small end of the cone, push tubing through all the way to the top and have plenty hanging outside, then tie the bottom of the cone to the base lathe wire.

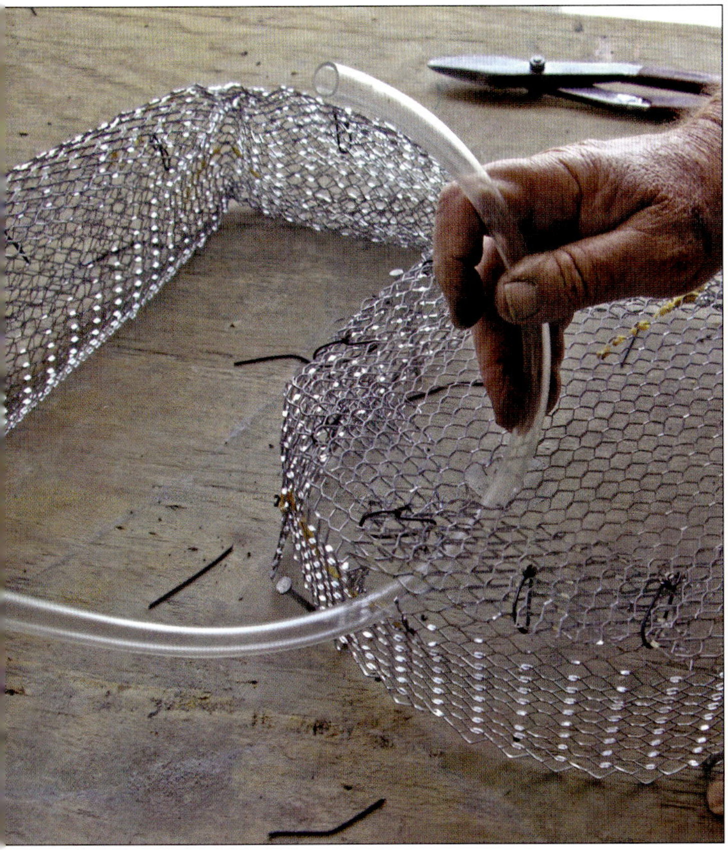

Push tubing through both holes as in photo.

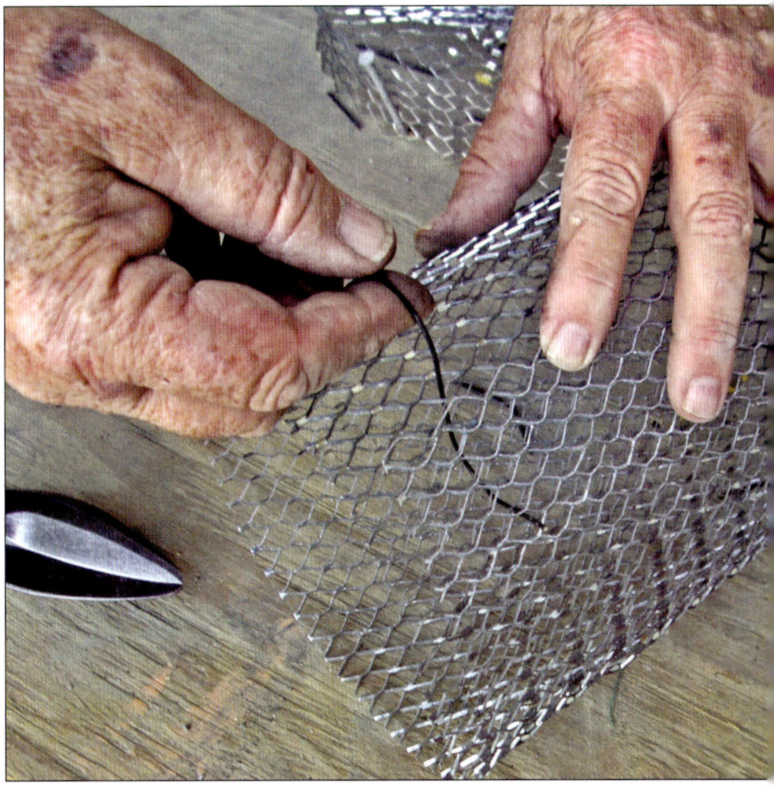

Cut another piece of lathe wire approximately 8" x 32" roll, overlap, and tie.

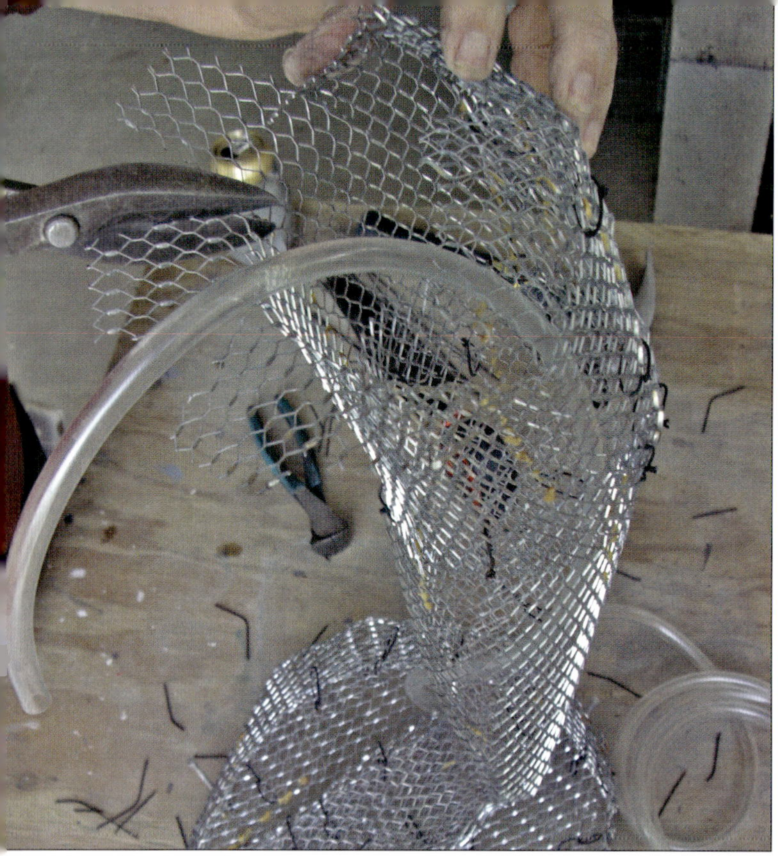

Take this piece of wire, try to overlap it, even just an inch on the top of the other lathe wire. What you are trying to do is make a curve in the dolphin's body towards the front. Make some 2" cuts around the top of the lathe and bend them outward.

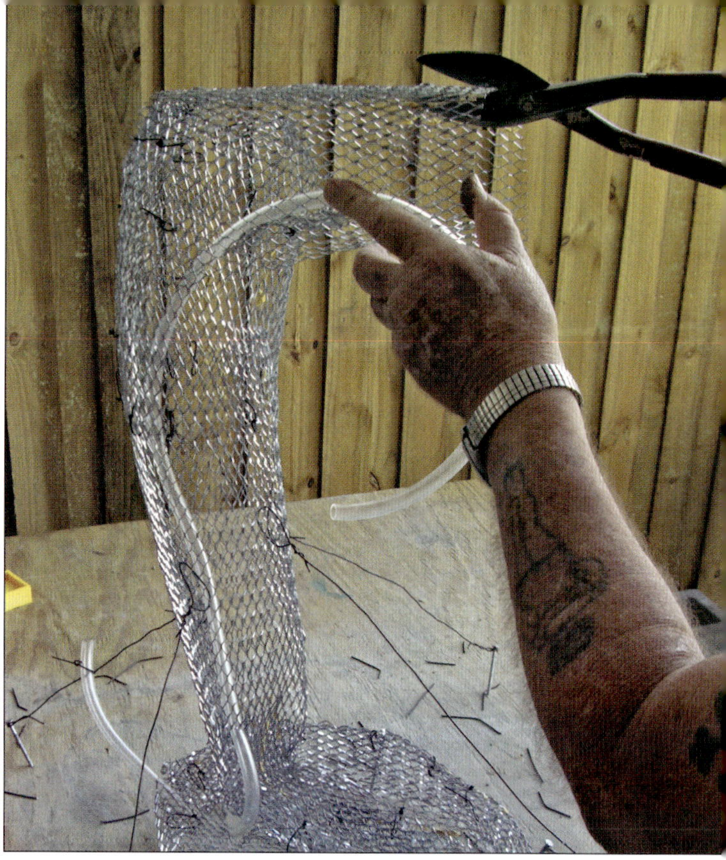

It is time to do something with his head. Cut a piece of lathe wire approximately 6" x 8". Roll it in a cone shape, then put it against the other lathe that is bent outward; now bend the wire together and tie. Make a couple of cuts where the nose of the dolphin will be, bend this lathe together and tie, making sure you do not put a tie around the tube because you will kink it and no water will flow through.

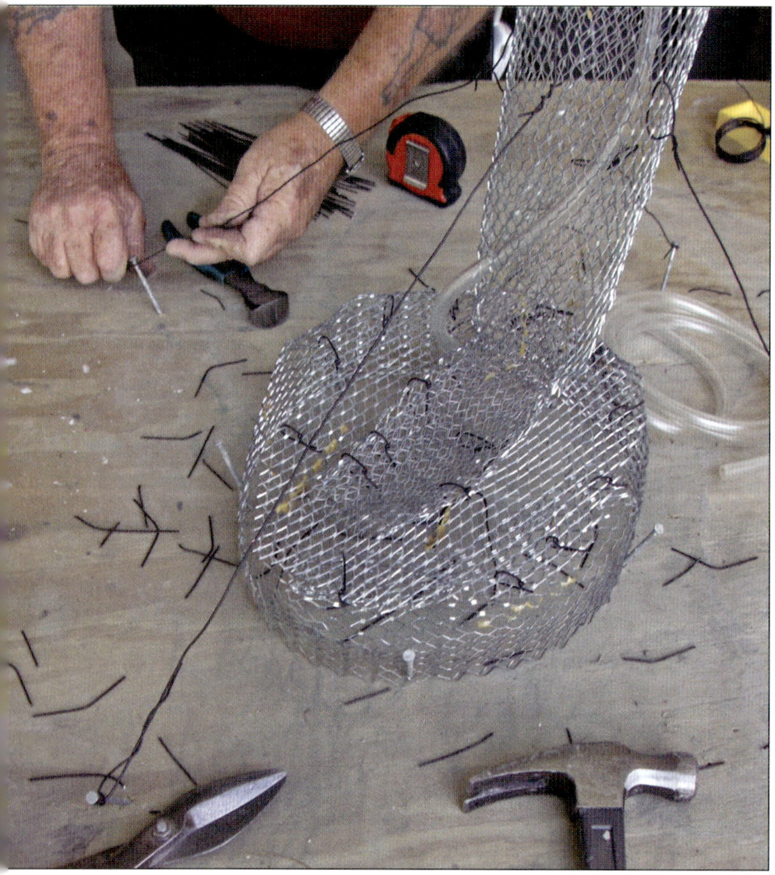

It is time to put four supporting wires on the dolphin.

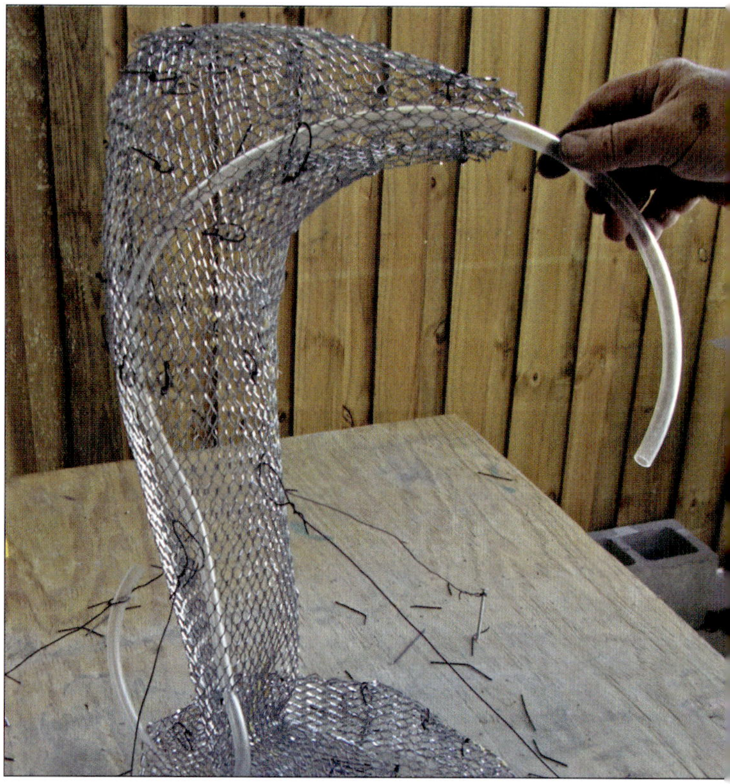

Just making sure there are no ties that are squeezing the tube.

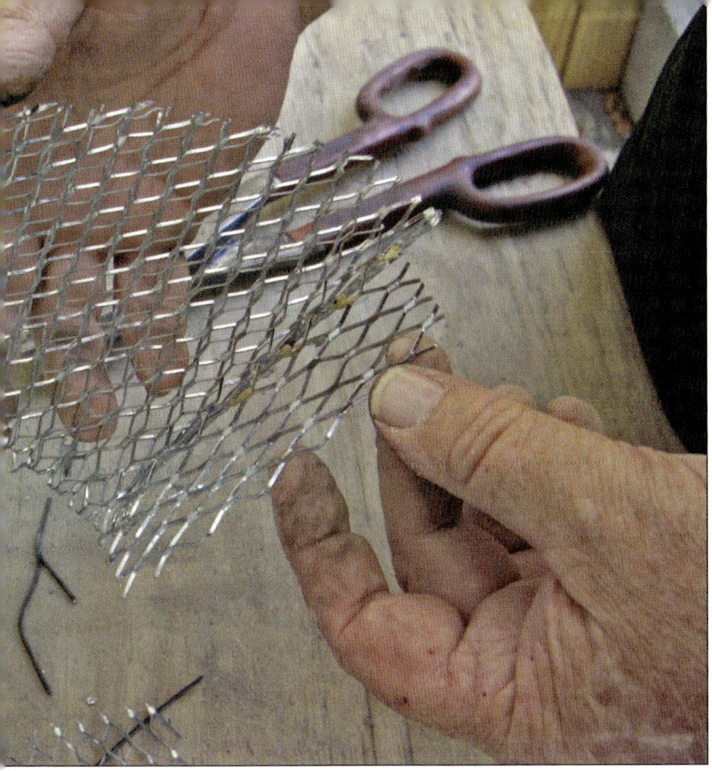

We will make the dolphin's fins, starting with the pectoral fin. Start by cutting a piece of lathe that is square-shaped. What we want here is double lathe wire. You can approximate the size of the square lathe, then bend it in half, and cut to shape with shears as in the photo.

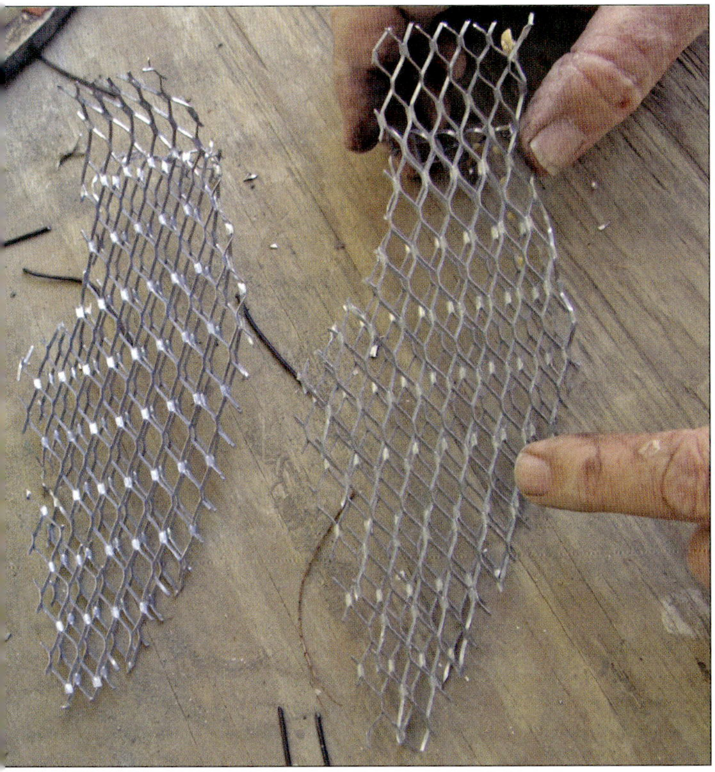

I am using the same procedure on the two side fins. After bending the lathe together, use shears to shape the lathe like dolphin fins and at the bottom of the fin that goes to the body, bend the lathe out about an inch. This is for tying to the body. For these fins, you can put one or two ties to keep the double lathe together.

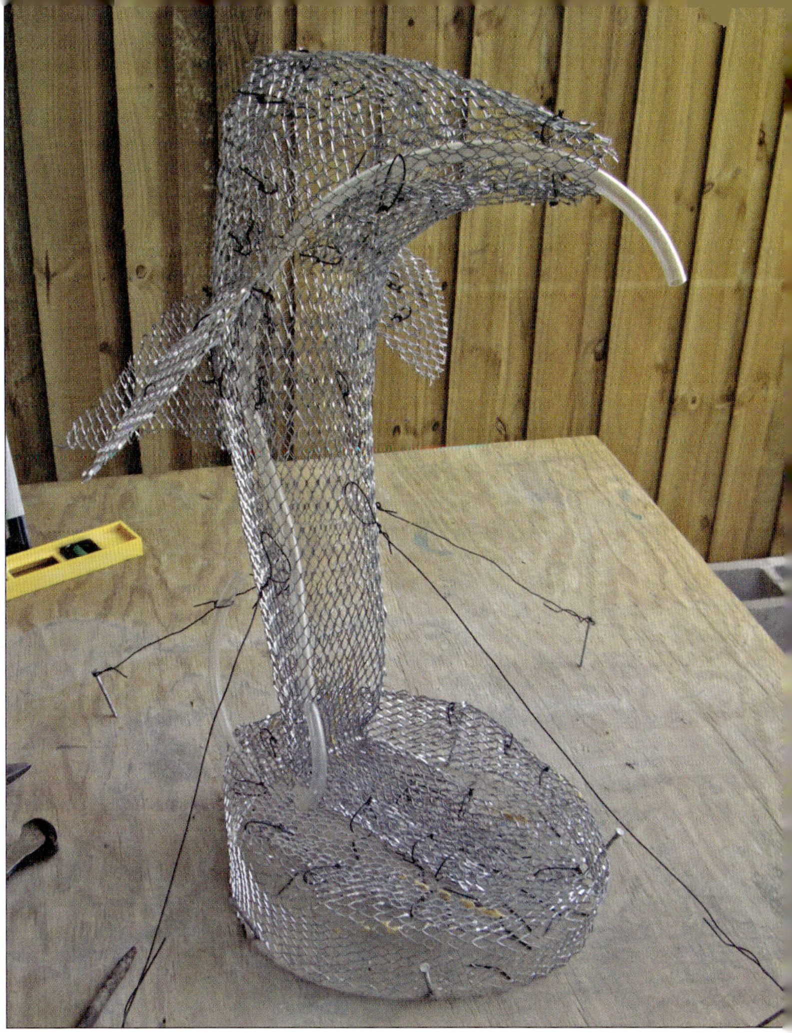

Ready for mud.

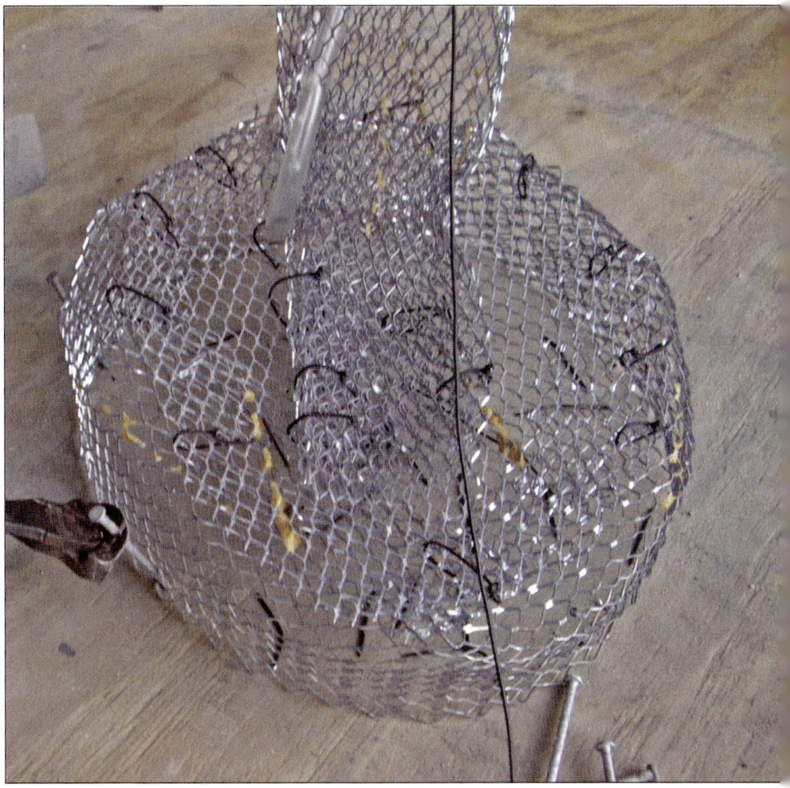

You can now pull nails out from around the base.

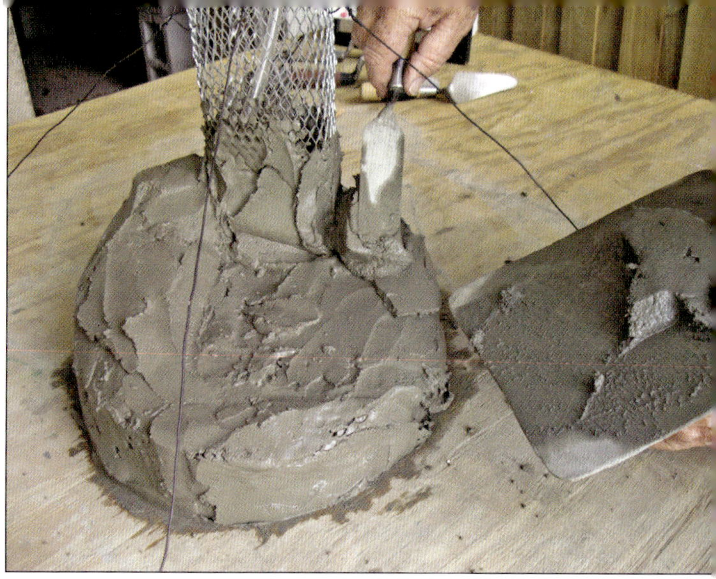

The base is solid. The bottom of the dolphin is solid approximately 3" to 4" up and now I'm building up the waves.

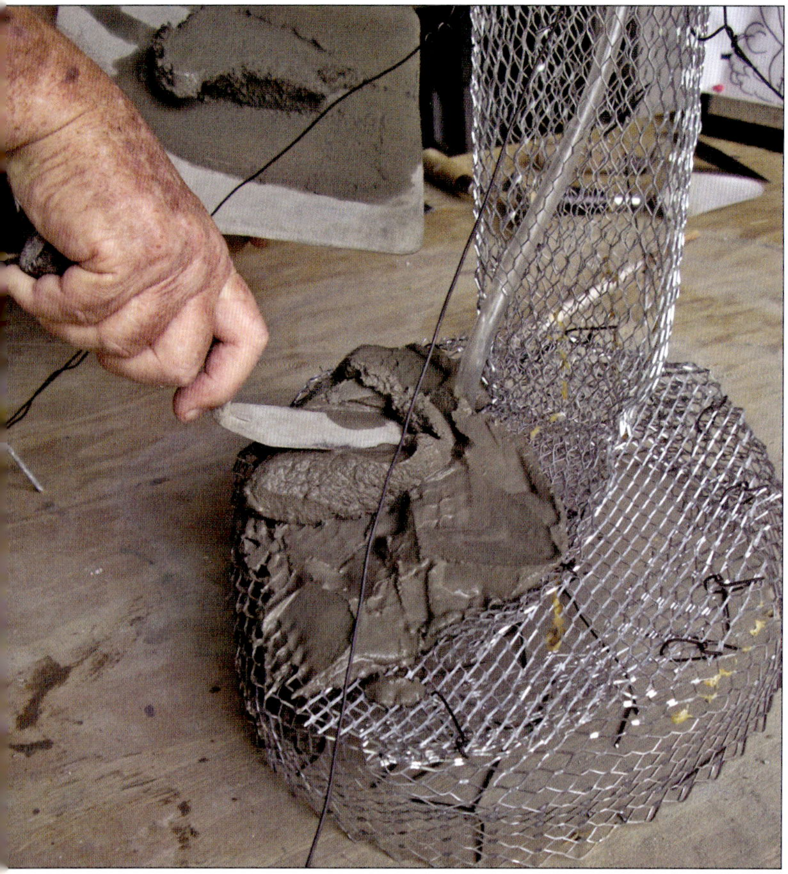

Working around the base, push mud through, lots of it. We are trying to make the base solid.

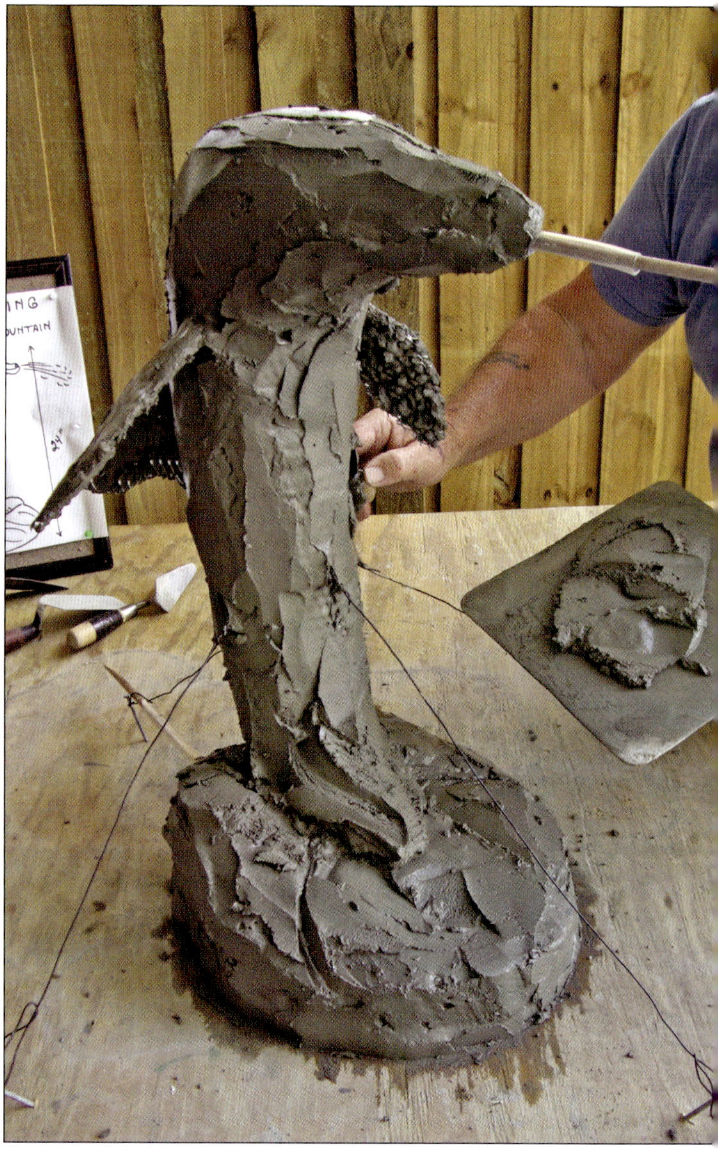

I have taken two pieces of 1/4" wooden dowels and placed inside the tube coming out of his mouth and the rear of the base. This is for keeping the tube straight until the mud sets.

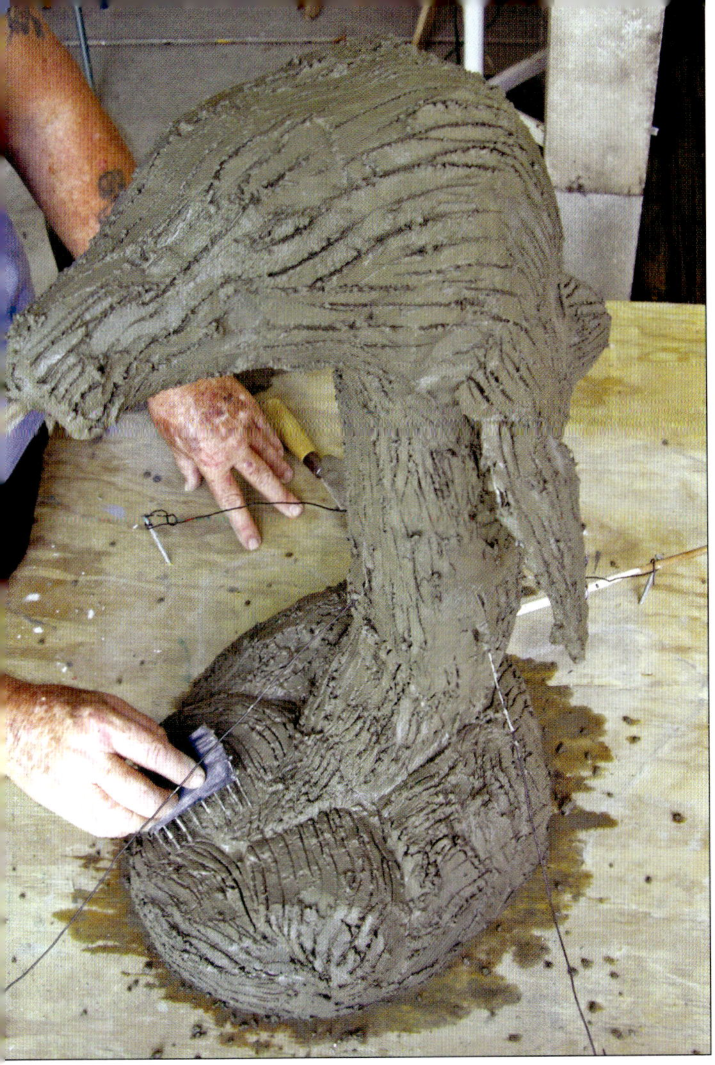

After building out some of the body and fins, it is time to use your scratch wire all over the sculpture.

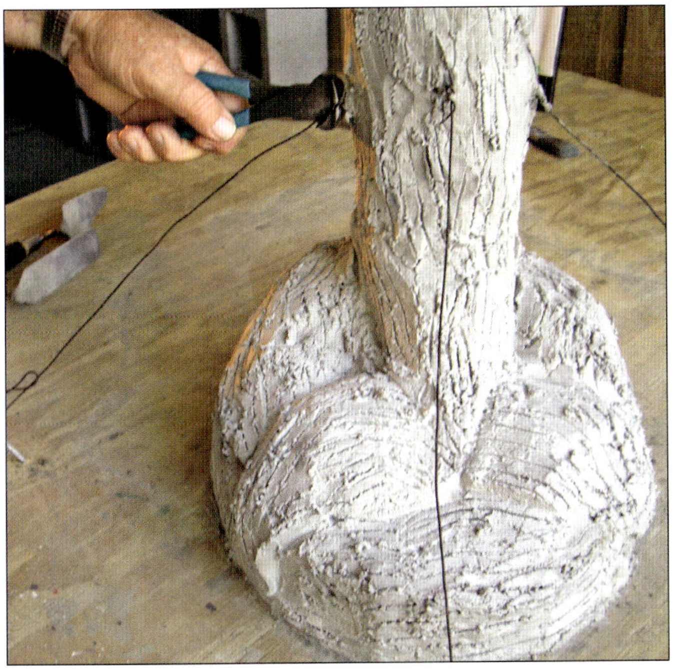

Here we go again! When you mix concrete it is soft, in one day it is hard and on the second day it is five times as hard. It is time to cut off the supporting wires.

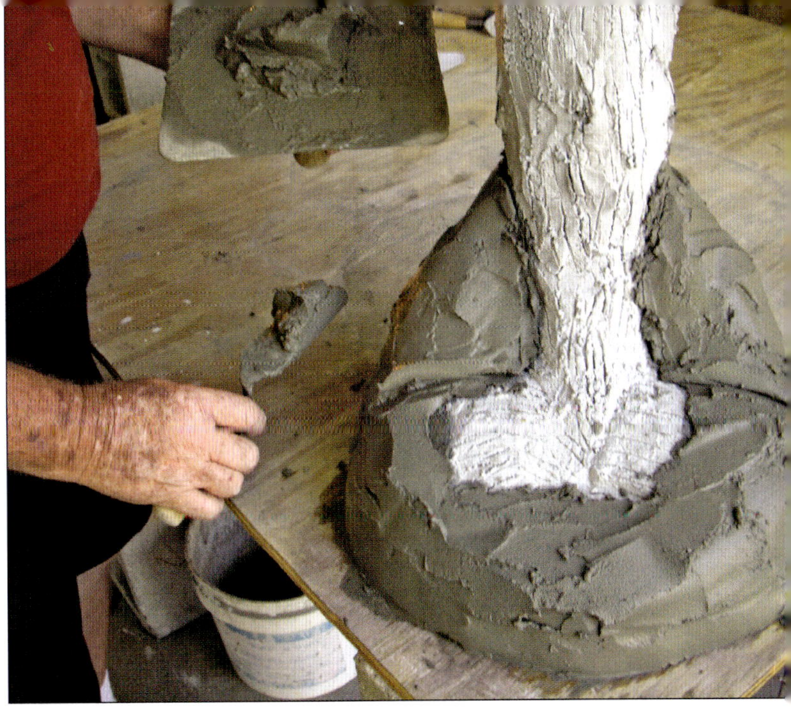

I'm starting at the bottom, building out the base and the waves.

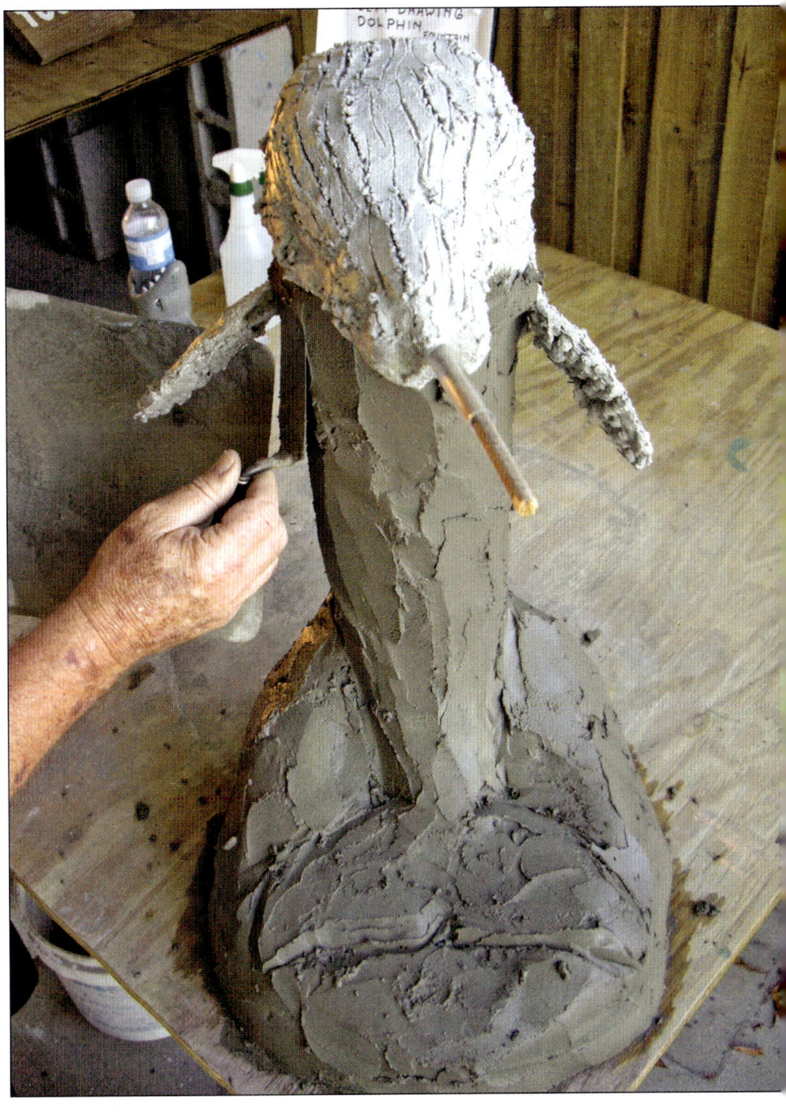

Bringing out the body from the bottom up starting with the fins.

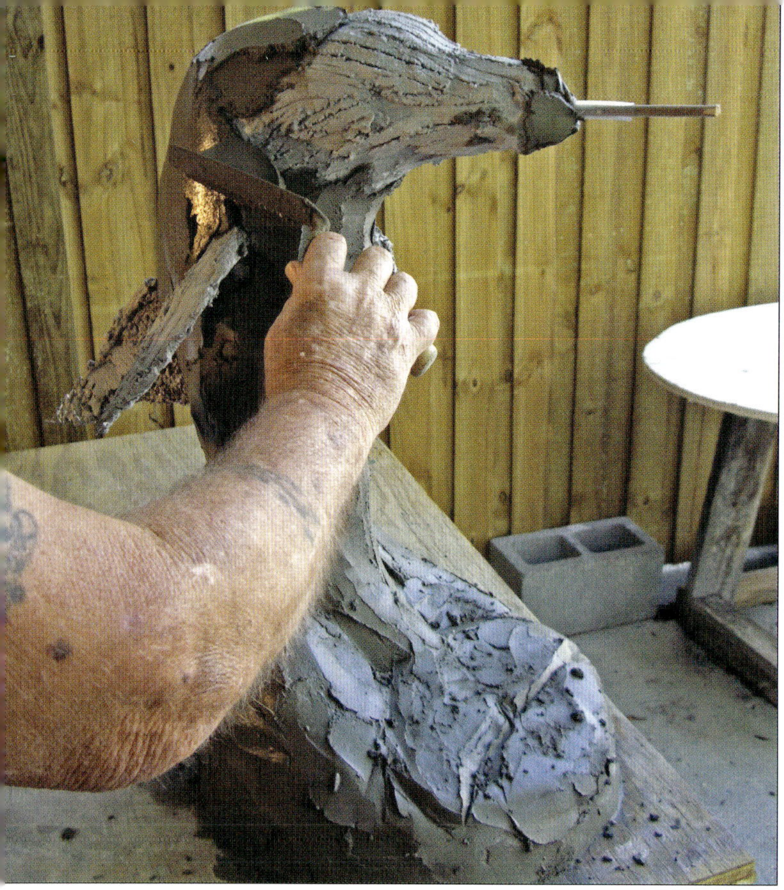

Building out the upper area of the dolphin.

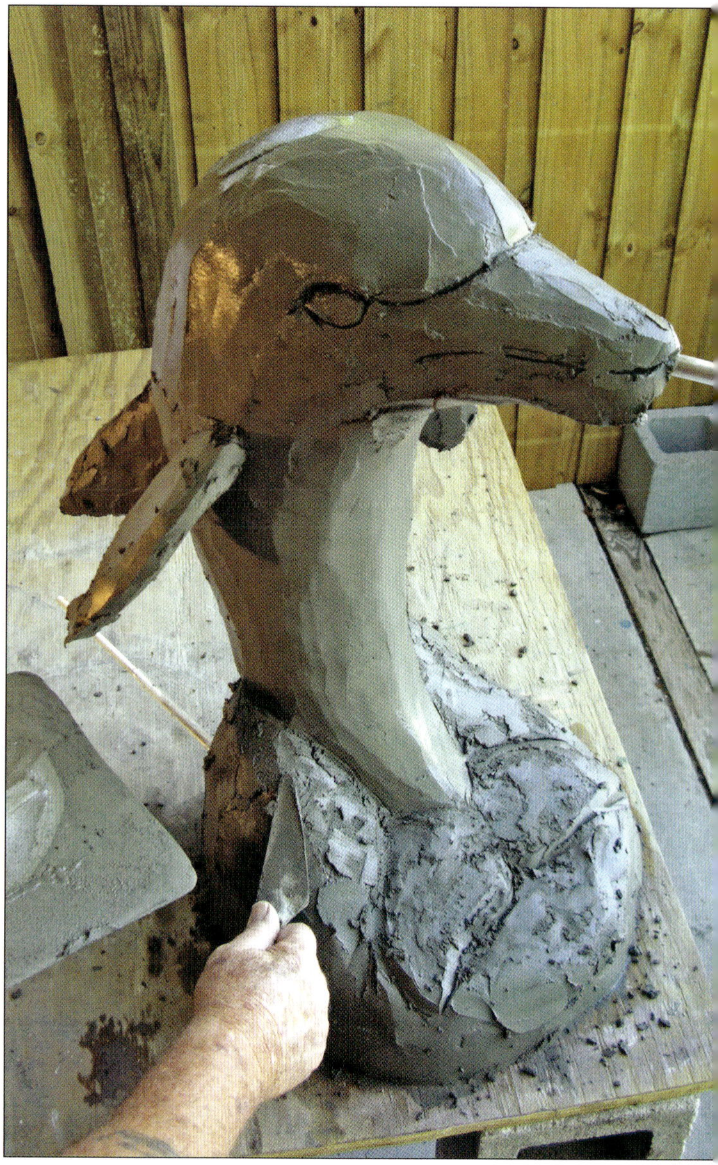

Head and body are starting to look like something.

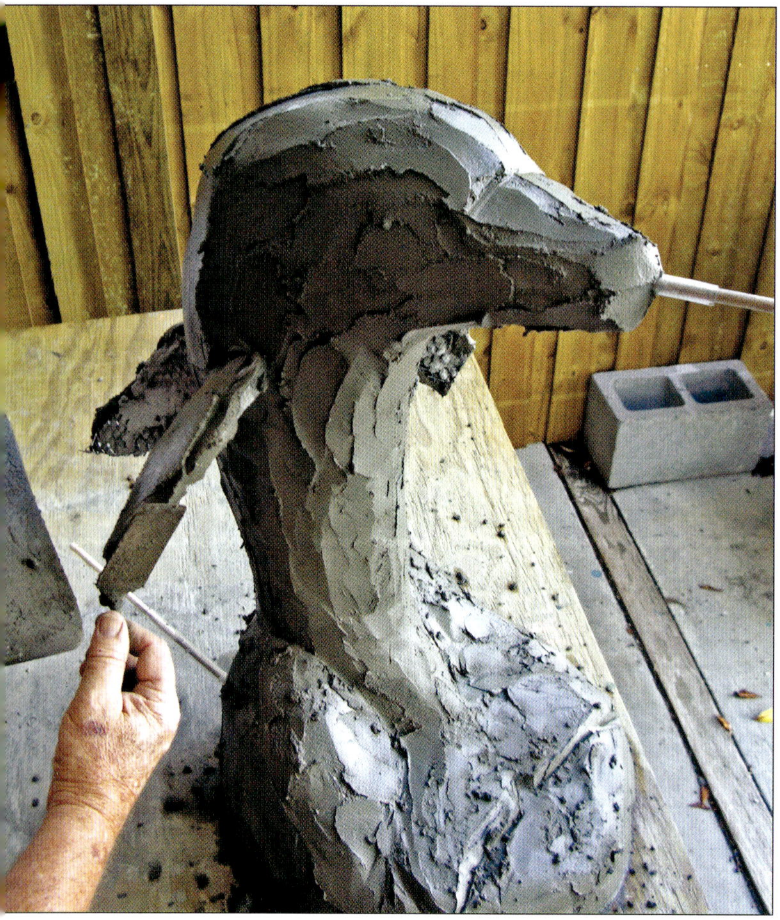

Working on the fins.

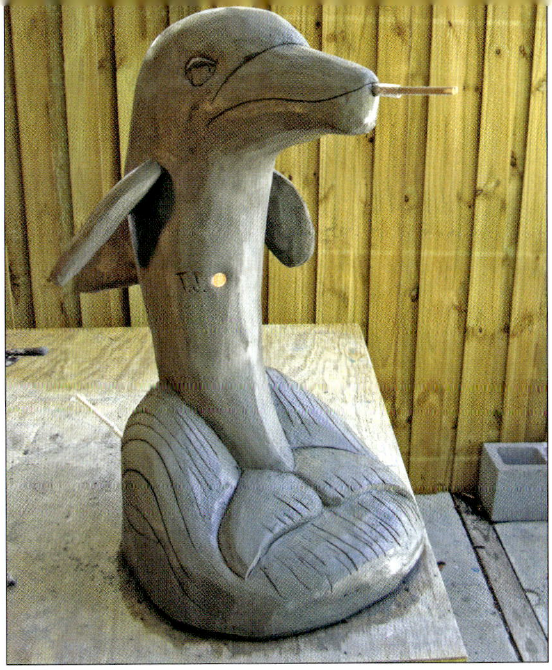

Finished mudding. I think it looks pretty good. If you have trouble making any sculpture smooth, there is another way to have a nice looking statue. Just take a damp sponge, when the concrete is set and you put your finger against it, it does not leave an impression. It is time to sponge float. Using circular motions with a damp sponge, what you are doing is bringing the sand up to the top and creating a sand finish.

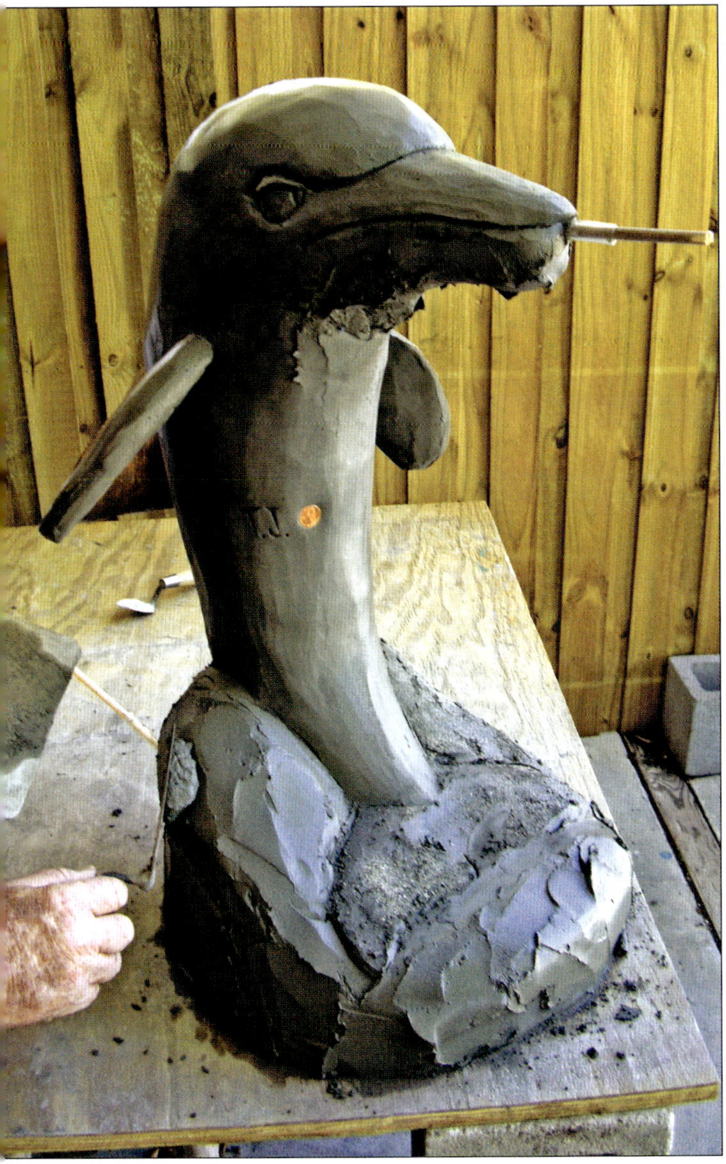

Initials and penny for luck.

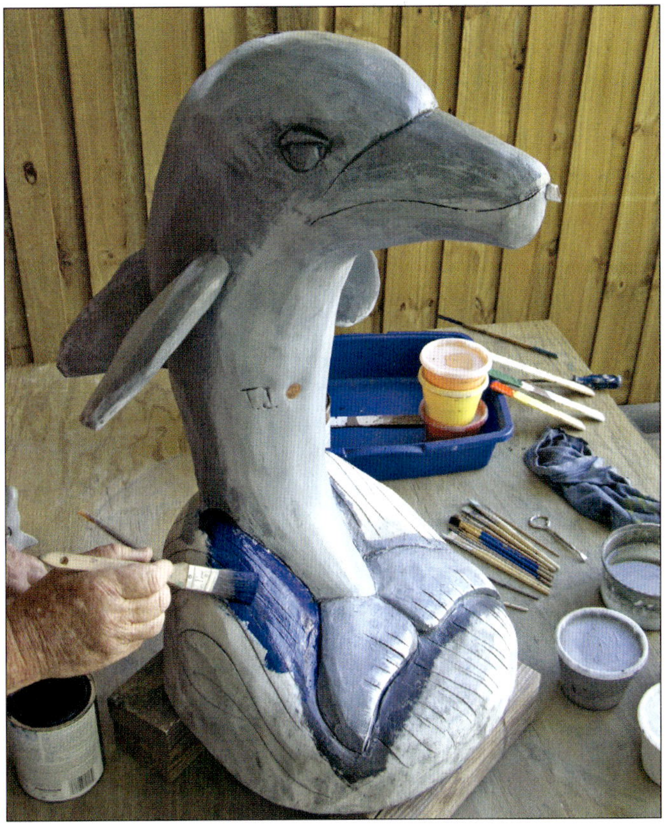

Take dowels out of the tubing. Using a sharp box knife, cut tubing at an angle, where it comes out of the mouth. The angle cut is for water to drip off without running down the body of the dolphin. Leave the tubing coming out of the base as long as you want.

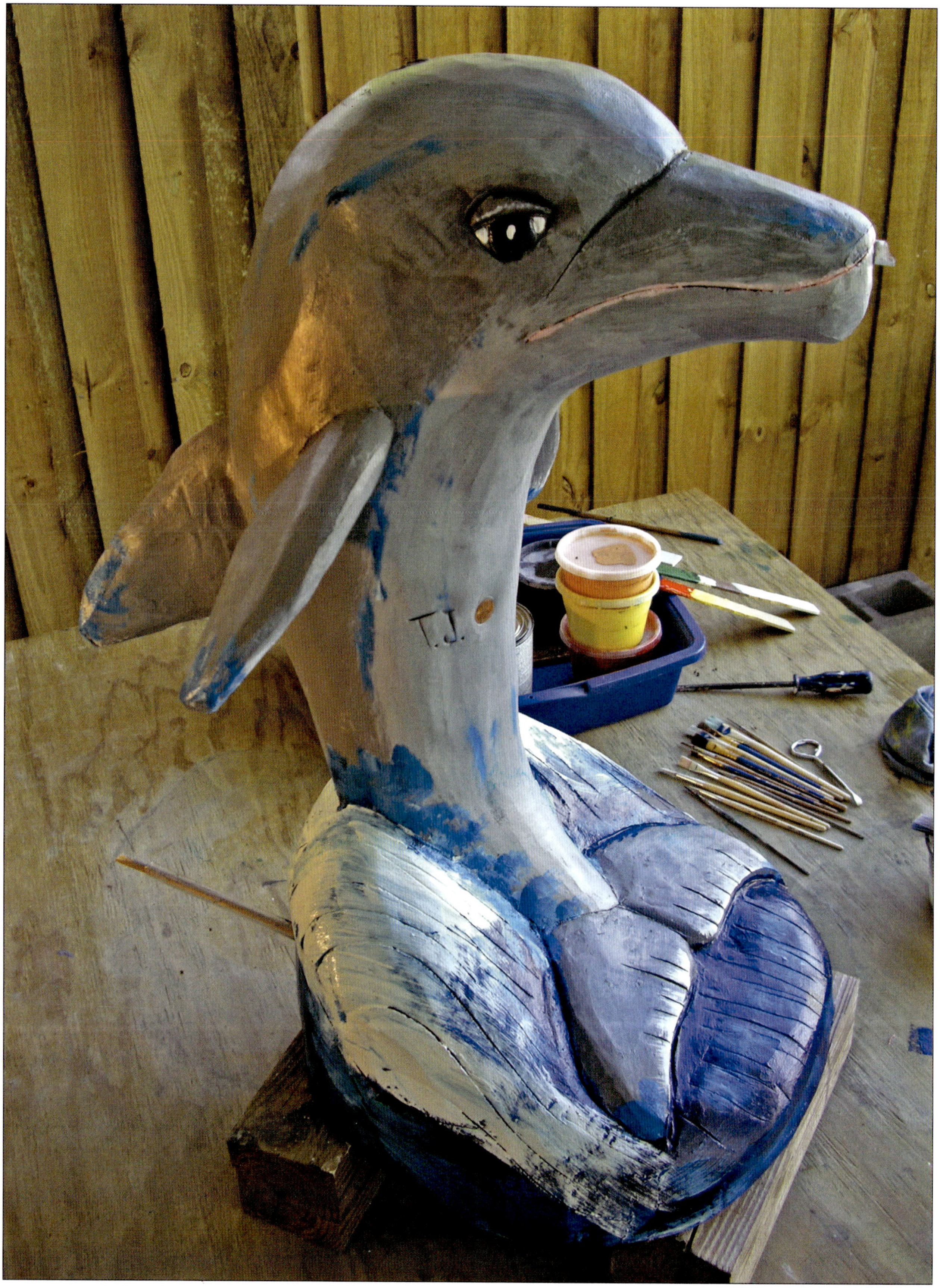

Dolphin Fountain all painted and the fountain is finished.